CONTENTS

Abbreviations		2
In Memory of the Late Brigadier General Ahmad Sadik Rushdie Al-Astrabadi		3
1	Ba'ath Setting the Stage	4
2	October 1973 Arab-Israeli War	7
3	Second Iraqi-Kurdish War, 1974–1975	21
4	Dulaymi's Troubled Tenure	27
5	Sha'ban's First Tour	32
6	The Rise of Saddam	40
7	IrAF of 1980	46
8	Echo of Qadessiya	56
References		71
Bibliography		71
Notes		73
About the Authors		78

NOTE

In order to simplify the use of this book, all names, locations and geographic designations are as provided in *The Times Comprehensive Atlas of The World*, or other traditionally accepted major sources of reference, as of the time of described events. Similarly, Arabic names are romanised and transcripted rather than transliterated. For example: the definite article al- before words starting with 'sun letters' is given as pronounced instead of simply as al- (which is the usual practice for non-Arabic speakers in most English-language literature and media). For the reasons of space, ranges – which are usually measured in feet and nautical miles in international aeronautics – are cited in metric measurements only.

Helion & Company Limited
Unit 8 Amherst Business Centre, Budbrooke Road, Warwick CV34 5WE, England
Tel. 01926 499 619
Email: info@helion.co.uk   Website: www.helion.co.uk   Twitter: @helionbooks   Visit our blog http://blog.helion.co.uk/

Published by Helion & Company 2022
Designed and typeset by Farr out Publications, Wokingham, Berkshire
Cover designed by Paul Hewitt, Battlefield Design (www.battlefield-design.co.uk)

Text © Milos Sipos and Tom Cooper 2022
Photographs © as individually credited
Colour profiles © Tom Cooper 2022
Maps and diagrams drawn by George Anderson © Helion & Company 2022 and Tom Cooper 2022

Every reasonable effort has been made to trace copyright holders and to obtain their permission for the use of copyright material. The author and publisher apologise for any errors or omissions in this work, and would be grateful if notified of any corrections that should be incorporated in future reprints or editions of this book.

ISBN 978-1-914377-17-4

British Library Cataloguing-in-Publication Data.
A catalogue record for this book is available from the British Library.

All rights reserved. No part of this publication may be reproduced, stored in a retrieval system, or transmitted, in any form, or by any means, electronic, mechanical, photocopying, recording or otherwise, without the express written consent of Helion & Company Limited.

For details of other military history titles published by Helion & Company Limited contact the above address, or visit our website: http://www.helion.co.uk. We always welcome receiving book proposals from prospective authors.

ABBREVIATIONS

AAA	anti-aircraft artillery
AAM	air-to-air missile
AB	air base
ADAR	*autodirector anti-radar* (seeker head of the AS.37 Baz-AR anti-radar missile)
ADOC	Air Defence Operations Centre (C3 or operational HQ of the IrAF)
AEW	airborne early warning
AMD-BA	Avions Marcel Dassault – Bréguet Aviation
An	Antonov (the design bureau led by Oleg Antonov)
AOI	(The) Arab Organisation for Industrialisation
ARM	anti-radar missile
ASCC	Air Standardisation Coordinating Committee
ATMS	automated tactical management system
AWACS	airborne early warning and control system
AWACW	Airborne Warning and Control Wing (USAF)
BAC	British Airspace Corporation (later BAe Warton)
BAe	British Aerospace
BAI	battlefield interdiction
BND	*Bundesnachrichtendienst* (Germany)
BVR	beyond visual range
C3	command, control and communication
CAP	combat air patrol
CAS	close air support
Chaff	reflective jamming device made up of thin, lightweight metallic strips, cut to one-half length of the target's radar wavelength
CO	commanding officer
COIN	counter-insurgency
COMINT	communications intelligence
CSB	Combat Support Base (major IRIAA base)
CTL	considered total loss
DGA	*Délégation Générale pour l'Armement* (General Delegation for Armament, former DMA renamed in 1977)
DIA	*Direction des Affaires Internationales* (Direction of International Affairs, part of the DMA, France)
DIA	Defense Intelligence Agency (USA)
DMA	*Délégation Ministérielle pour l'Armement* (Ministerial Delegation for Armament, France, 1965-1977)
DMI	Directorate of Military Industries
EAF	Egyptian Air Force
ECM	electronic countermeasures
ELINT	electronic intelligence
FCS	fire-control system
Flare	pyrotechnic device released from an aircraft causing an infra-red homing missile or target tracker to follow it rather than the aircraft
FLS	Flight Leaders School
FOB	forward operating base (also 'dispersal site')
GMID	General Military Intelligence Directorate (Iraq)
HAWK	Homing-All-The-Way-Killer (US-made SAM)
HPIR	High Power Illuminator (Doppler) Radar (X-band tracking radar guiding MIM-23 HAWK SAMs)
IADS	integrated air defence system
IAP	international airport
ICP	Iraqi Communist Party
IFR	in-flight refuelling
I-HAWK	Improved Homing-All-The-Way Killer (US-made SAM)
Il	Ilyushin (the design bureau led by Sergey Vladimirovich Ilyushin, also known as OKB-39)
IIAF	Imperial Iranian Air Force
INA	Iraq News Agency
IrAAC	Iraq Army Aviation Corps
IrAF	Iraq Air Force (official designation since 1958)
IRGC	Islamic Revolutionary Guards Corps
IRIAA	Islamic Republic of Iran Army Aviation
IRIAF	Islamic Republic of Iran Air Force
IRIN	Islamic Republic of Iran Navy
KDPI	Democratic Party of Iranian Kurdistan
KGB	*Komitet Gosudarstvennoy Bezopasnosti* (Committee for State Security, USSR)
Lock-on	continuous and automatic tracking of a target by the radar and the FCS
MANPAD	man-portable air-defence system (shoulder-fired anti-aircraft missile)
MIC	Military Industrialisation Commission (later DMI, Iraq)
MiG	Mikoyan i Gurevich (the design bureau led by Artyom Ivanovich Mikoyan and Mikhail Iosifovich Gurevich)
NATO	North Atlantic Treaty Organisation
nav/attack	navigational and attack (avionics suite)
NSA	National Security Agency (USA)
OCU	Operational Conversion Unit (IrAF)
ONI	Office of Naval Intelligence (US Navy)
PLO	Palestinian Liberation Organisation
RAF	Royal Air Force (United Kingdom)
RIO	radar intercept officer (back seater in the F-4E Phantom II and F-14A Tomcat)
RSAF	Royal Saudi Air Force
RWR	radar warning receiver
SAM	surface-to-air missile
SAR	search and rescue
SARH	Semi-Active Radar Homing (missile guidance system)
SEPECAT	*Société Européene de Production de l'avion Ecole de Combat et d'Appui Tactique* (European Company for Production of Training, Combat, and Tactical Support Aircraft; joint venture of Bréguet and BAC)
SIGINT	signals intelligence
Su	Sukhoi (the design bureau led by Pavel Ossipovich Sukhoi)
SyAAF	Syrian Arab Air Force
TAB/TFB	Tactical Air Base/Tactical Fighter Base (IRIAF)
TWT	travelling wave tube
UAE	United Arab Emirates
USAF	United States Air Force
USN	US Navy
USSR	Union of Soviet Socialist Republics (also 'Soviet Union')
VLCC	very large crude carrier ('supertanker' with a hull length of more than 330 metres)
VVS	*Voyenno-Vozdushnye Sily* (Soviet Air Force)

Announcement of the death of Brigadier General Ahmad Sadik Rushdie al-Astrabadi, published in Baghdad in September 2019. (via Tom Cooper)

In Memory of the Late Brigadier General Ahmad Sadik Rushdie Al-Astrabadi

This book summarises more than three decades of research into the Iraqi Air Force (IrAF). It might sound unusual, but the work on it began long before the work on Volume 1 of this mini-series was initiated. Its origins date back to the 1980s, when – while still young enthusiasts – we began exploring air warfare between Iran and Iraq, and the related lessons. Many of its parts came into being 'on their own', as we continued collecting details about that long and bitter conflict, but the final impetus was provided by the rare opportunity to extensively interview and work with Brigadier General Ahmad Sadik Rushdie al-Astrabadi – retired officer of the Intelligence Directorate of the Iraqi Air Force – in the course of four meetings in Syria, between 2005 and 2007. Indeed, a large part of what was merged into Volume 2 came into being through cooperation with Ahmad, and he can be considered not only the crucial source for most of this information, but its co-author.

The first volume of this mini-series covered the first 40 years of history of the IrAF, from the background and establishment of Iraq in 1920–1921, to the establishment of the air force in 1931, through several wars in which it participated, and the build-up to 1970. Volume 2 covers a much shorter period of time from 1970 until 22 September 1980, the day on which Iraq invaded Iran. Our original intention was to close this project with Volume 2, that is, cover all the 30 years of this once-proud service up to its demise in 2003. However, it turned out that the amount of information left behind by Ahmad, and the amount of information collected over the last decade enable a much deeper, more detailed history of the IrAF than expected.

Furthermore, the 1970s in particular were – acquisitions and operations-wise – the second most intensive period in this history of the IrAF (right before the eight-year-long war with Iran). Within just a few years, the IrAF grew at an unprecedented scale, tripling its total manpower, while massive investment into the future aircraft and equipment of those times have shaped its appearance, capacity and capabilities right to the end in 2003. This is why this volume spans 'just' 10 years.

ASCC/NATO-CODENAMES FOR SOVIET-MADE AIR-TO-AIR, AIR-TO-SURFACE, AND SURFACE-TO-AIR ORDNANCE MENTIONED IN THIS BOOK

The IrAF of the period 1970–1980 was shaped not only by wars and combat experiences, but also the acquisition of new equipment – and especially equipment made in the Union of Soviet Socialist Republics (USSR, colloquially 'Soviet Union'). Some of the equipment in question is still rather unknown in the English-language world, segments of this book are likely to become hard to follow. For this reason, Table 1 provides a review of Soviet designations, their ASCC/NATO-codenames, and other important details.

Table 1: ASCC/NATO-Codenames for Soviet-made Air-to-Air, Air-to-Surface, and Surface-to-Air Ordnance

Soviet Designation	ASCC/NATO-Codename	Guidance & Notes
Air-to-Air Missiles		
R-3S	AA-2A Atoll	IR-homing
R-13M	AA-2D Atoll	IR-homing
R-23R	AA-7A Apex	SARH
R-23T	AA-7B Apex	IR-homing
R-27R	AA-10A Alamo	SARH
R-40RD	AA-6C Acrid	SARH
R-40TD	AA-6D Acrid	IR-homing
R-60MK	AA-8 Aphid	IR-homing
Air-to-Ground Missiles		
Kh-66 Grom		radio-command guidance
Kh-23/Kh-23M	AS-7 Kerry	radio-command (Kh-23) or semi-automatic radio-command guidance (Kh-23M)
Kh-28	AS-9 Kyle	passive radar homing
Kh-25M	AS-10 Karen	radio-command guidance
Kh-25MP	AS-12 Kegler	passive radar homing
Kh-29L	AS-14 Kedge	laser guidance
Kh-58E	AS-11 Kilter	passive radar homing
Surface-to-Air Missiles		
S-75 Dvina/SA-75 Desna/S-75M Volkhov	SA-2 Guideline	radio-command guidance
S-125 Neva/Pechora	SA-3 Goa	radio-command guidance
2K12 Kub/Kvadrat	SA-6 Gainful	command guidance with terminal SARH
9K32M Strela-2M	SA-7 Grail	IR-homing
9K33 Osa/Romb	SA-8 Gecko	command to line-of-sight
9K31 Strela-1	SA-9 Gaskin	IR-homing
9K35 Strela-10	SA-13 Gopher	IR-homing

1
BA'ATH SETTING THE STAGE

On 17 July 1968, the leadership of the Ba'ath Party of Iraq, supported by officers of the General Military Intelligence Directorate (GMID; colloquially 'Mukhabarat'), the Republican Guard, the IrAF, and in cooperation with the Iraqi Communist Party (ICP), staged a coup d'état in Baghdad and toppled the government of President Abd ar-Rahman Arif. A new government established itself in power in the form of the Revolutionary Command Council (RCC), presided over by Major General Ahmed Hassan al-Bakr, several officers involved in the coup, and four Kurdish representatives. When it turned out that the RCC was dysfunctional, the military wing of the Ba'ath Party staged a new coup on 30 July 1968. Bakr retained power as the President and Prime Minister in charge of a 26-strong cabinet; however, the actual executive and legislative power was concentrated in the RCC – also presided over by Bakr and consisting of four other military officers and Ba'ath members – which ruled by decree. In addition to Ahmed Hassan al-Bakr, these were Salih Mahdi Ammash, Hardan Abd ar-Rhefal at-Tikriti (Chief of Staff IrAF, and the new Minister of Defence), Sa'aydoon Ghaydan and Hammad Shihab at-Tikriti. Contrary to what might have been expected, Bakr, a popular military officer, then sided with the civilian wing of the party: he forged a tactical alliance with his younger cousin Hussein Abd al-Majid at-Tikriti – nicknamed 'Saddam' (literally 'Hammer' in Arabic; colloquially 'Saddam' or 'Saddam Hussein') – who was in charge of the internal security of the party, and assigned him the role of eliminating political rivals outside the Ba'ath.[1]

TIKRIT VERSUS TIKRIT

Born in a small village outside Tikrit, Saddam grew up in an atmosphere of corruption, mistrust, patronisation, and brutal force, all of which made him mercilessly violent – which helped him work himself up the ranks of the Ba'ath through the 1960s, when he liquidated most of his opponents and rivals. Exiled in Egypt, Saddam played no role in the two coups of 1963; he returned only after the second, in time to find out that the Ba'ath Party Militia had been disbanded. After serving a sentence of two years in 1964–1966 for the attempted assassination of President Arif, he re-emerged in time to take part in the two bloodless coups of 1968. Proven as a skilled organiser and brutal executioner, Saddam then planned and carried out a purge of political opponents in the armed forces, and 2,000 Nasserists, Syrian- or Communist sympathisers were forced to leave the service over the following 12 months. They were replaced by about 3,000 Ba'athists, the mass of whom were actually civilians with minimal military training, and a number of officers either originating from Tikrit, or other Sunni Arab tribes of north-western Iraq. He thus built his reputation to the point where, in November 1969, Saddam was appointed the Vice-Chairman of the RCC, and thus formally al-Bakr's second in command. The two formed an unbeatable duet; while Bakr focused on controlling the armed forces and modernising the Iraqi economy, Hussein controlled the security apparatus and the Ba'ath's National Security Bureau. The combination proved sufficiently powerful to quickly put down any coup attempts.[2]

Once allied, Bakr and Saddam focused on removing their rivals from the RCC. Primary targets became Hardan at-Tikriti – a popular veteran of the Palestine War in 1948, but widely perceived as a 'nominal Ba'athist' – and Salih Mahdi Ammash, both of whom retained a strong power base within the armed forces. In a perfect example of skilful political manoeuvring, Bakr disconnected both of them by merging the portfolio of the Minister of Defence with that of the Prime Minister, and then assuming both positions, in turn disconnecting Hardan from his followers. The showdown came in September 1970, when Hardan at-Tikriti argued against the idea of attacking Jordan in support of the Palestinians that found themselves at odds with the local army loyal to King Hussein. Complaining about Hardan's 'inaction', Bakr and Saddam had him – and everybody affiliated – purged from the armed forces: Hardan was exiled to Algeria and replaced by Brigadier General Hussein Hiyawi Hamash at-Tikriti as the Chief of Staff IrAF. By 1971, Bakr and Saddam had removed Ammash too, leaving the former as the only high-ranking military officer with a strong support base in the armed forces *and* in a position of executive power.[3]

THE END OF COLONIAL TIMES
Always concerned with widening their power base and mobilising popular support, next Bakr and Saddam staged the first of two major coups on the international scene. Following two months of negotiations in Baghdad and Moscow, on 9 April 1972, Iraq and the Soviet Union signed a 15-year friendship treaty; an act that became even more significant when Egyptian President Anwar el-Sadat expelled most of his Soviet military advisors only a few months later. Iraq thus secured the status of a major Soviet partner in the Middle East and, unsurprisingly, the leadership in the Kremlin even granted permission for the IrAF to place an order for 14 Tupolev Tu-22 supersonic bombers, and brand-new MiG-23 swing-wing interceptors. To pay for such acquisitions

Hardan Abd ar-Rhefal at-Tikriti, Chief of Staff, IrAF, and Minister of Defence of Iraq, in the period 1968-1970, seen opening the Baghdad International Exhibition in 1969. Notable behind him is the contemporary Vice-President of Iraq, Saddam Hussein at-Tikriti. (via Ali Tobchi)

and to further improve their standing with the public, Bakr and Saddam then moved against the metaphorical thorn which had been in the side of Iraq as a nation ever since inception: the foreign control of the country's oil and gas resources, at the time estimated to be about 50 percent of known worldwide reserves. Relations between Baghdad and the Iraqi Petroleum Company (IPC) – now jointly owned by British, US, French and Dutch companies – had

A plaque at the entrance to the North Rumaila oilfield, declaring it a result of the Iraqi-Soviet agreement from 4 June 1969, which was opened on 19 July 1972. The issue became a thorn in the side of the IPC and several Western governments. (INA/TASS)

Alexey Kosygin (left) and Ahmed Hassan al-Bakr signing the Iraqi-Soviet Treaty of Friendship and Cooperation, on 9 April 1972. (INA/TASS)

never been friendly. However, they turned into open enmity when Baghdad nationalised the IPC's unexploited territories in 1961, and then began developing these – foremost the gigantic North Rumaila oilfield – in cooperation with the USSR. As arrogant as ever, the directors of the IPC demanded financial compensation: when years-long negotiations ended in a stalemate, the company cut the exports from its oilfields in the Kirkuk area by 50 percent in early 1972. After quickly organising new markets for Iraqi oil in Europe – primarily France and Italy – on 1 June 1972 Bakr and Saddam nationalised the entire Iraqi oil industry, including the IPC. To summarise the effects of this decision as 'dramatic' would be an understatement: not only was the government's decision widely celebrated in Iraq as 'the end of colonial times', but the Iraqi oil revenues jumped from about US$800 million in 1971 to about US$1.6 billion by the end of 1972, enabling Bakr and Saddam to develop one of the most modern public health systems with free hospital care for everyone, a welfare system for military servicemen and their families, subsidies for farmers, and the launch of a comprehensive national infrastructure campaign that made great and rapid progress in building roads and irrigation systems. Within a few years, electricity was brought not only to all of the cities, but most of the outlying areas, too: by 1973, the Iraqi economy was booming.

LOSS OF VETERANS

Despite successful economic development, the purges of Hardan at-Tikriti and Salih Mahdi Ammash and efforts to distance the military from politics with the help of the Saddam-led security system of the Ba'ath – and a network of his clan and family relations – caused much unrest within the armed forces. Conducted during periods of major expansion and particularly intensive exercises, the purges hit all three branches severely because of the resulting loss of dozens, if not hundreds of experienced officers – or officers fresh from training courses abroad. Unsurprisingly, political dissent remained, frequently causing unrest. In January 1970, retired Major General Abd al-Ghani ar-Rawi and active Colonel Salih Mahdi as-Samarrai led a contingent from the Republican Guards Brigade in an attack on the Republican Palace in Baghdad. The plot was uncovered and the column ambushed; while Rawi fled to Iran, 27 officers and other ranks involved were sentenced to death and executed. A few months later, in the wake of Hardan at-Tikriti being forced into exile, the IrAF then lost Captain Majid Turki al-Jumaily, the co-pilot of one of two Tu-16s that bombed Israel on 6 June 1967: he was arrested for 'conspiracy against the state', and executed.[4] A few days later, Samir Youssef Zainal, hero of the June 1967 War with Israel – where he had scored a confirmed kill against a SNACSO S.O.4050 Vautour fighter-bomber while flying a Hunter – was forced into exile. Another disgruntled veteran that left service around this time was Farhad Ibrahim, who defected to Ahwaz Airport in Iran flying a brand-new Sukhoi Su-7BMK fighter-bomber, on 18 August 1971, only weeks after returning from a training course in Czechoslovakia. Ibrahim parted from his formation during a routine training flight, and disappeared by flying at low altitude; an extensive search and rescue operation for him remained fruitless, and it was only once Tehran offered to return the aircraft to Iraq that Baghdad realised what had happened.

Captain Majid Turki al-Jumaily (foreground, right, looking at the camera), seen in front of one of No. 10 Squadron's Tu-16s in the late 1960s. (via Ali Tobchi)

While many officers and pilots of the IrAF were purged in 1968-1971, even more new pilots completed their basic training - and many of them were to play an important role over the next decade. Amongst them was the future Major-General Qais Rabie abd al-Rahman al-Obeidi, who was to serve on Su-20s and Su-22s, and took over as the CO No. 109 Squadron in 1981. (via Ali Tobchi)

Eventually, Lieutenant Muhammad Mudher al-Farhan was sent to Iran to pick up the Sukhoi.[5]

Disagreement was not limited to the Army and Air Force. In 1973, Nadhim Kazzar, a Shi'a but appointed by Saddam as the Director of the General Security Service, launched a coup attempt; this failed, but Kazzar took the Minister of Defence and Minister of Interior as hostages, and attempted to escape to Iran. He was intercepted and, after shooting both ministers, he was arrested, sentenced to death and executed. The Kazzar coup attempt prompted Saddam into taking over the control of the General Security Service. He appointed his younger half-brother, Barzan Ibrahim at-Tikriti as the head of the GMID, and another of his close associates, Taha Yassin Ramadan (a Kurdish Ba'athist) as the commander of the re-established Ba'ath Party Militia (now named the 'People's Army'). He also established another intelligence service in the form of the Directorate of Political Guidance, responsible for monitoring the armed forces. Henceforth, all major ammunition and fuel depots were under the control of party members, and all large movements of the armed forces required permission from Bakr and Saddam. Ultimately, the government was successful: it not only 'Ba'athificated' the military establishment, but also secured party control over all key security institutions in and around Baghdad.

2
OCTOBER 1973 ARAB-ISRAELI WAR

Through 1971–1973, the Iraqi Air Force continued pursuing multiple projects that had been launched in the 1966–1967 period. Its total manpower grew to around 9,500 by 1973, and it introduced to service additional MiG-21F-13s, MiG-21FLs, MiG-21PFMs and Sukhoi Su-7BMKs from the USSR. Through the second half of 1969, several additional arms deals were concluded with Prague and Moscow, which included deliveries of:

- 16 refurbished MiG-15bis, from Czechoslovakia in 1971
- followed by up to 22 MiG-15bis a year later
- 98 brand-new MiG-21M/MF interceptors

FORGOTTEN HERO

Perhaps the best – and also saddest – example of the victims of Saddam's persecution of officers within the ranks of the IrAF in 1970 was Samir Youssef Zainal. Graduate of the 10th Class at the Air Force College (1948), he was trained to fly British-made Hawker Hunters at RAF Chivenor in 1957–1958. As described in Volume 1, he earned himself the status of war hero for his involvement in the third air combat over airfield H-3 on 7 June 1967, when officially credited with an aerial victory against a SNCASO 1050 Vautour fighter-bomber of the Israeli Defense Force/Air Force (IDF/AF). Zainal continued a successful career with the IrAF during the late 1960s. By 1970, he was the commanding officer (CO) of No. 29 Squadron based at al-Taqaddum AB. During Saddam's purge of the air force in the aftermath of the expulsion of Hardan at-Tikriti, Zainal was deported to Libya, with 'behaviour outside established orders' as the official explanation: apparently, he ordered his ground crews to arm and refuel one of the Hunters without permission from his superiors. Zainal subsequently moved to Syria and joined the Syrian Arab Air Force (SyAAF). By 1973, he was commander of an air brigade including three Syrian and one Egyptian MiG-17 squadrons, headquartered at Mezzeh AB, in Damascus. Samir Youssef Zainal was killed in the line of duty, while leading his pilots in an attack on Israeli troops in the Mazra'at Beit Jann area, in south-western Syria, on 13 October 1973.

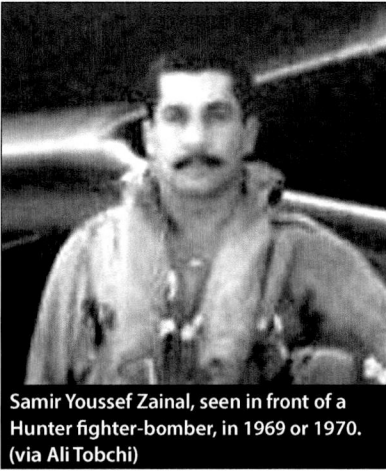

Samir Youssef Zainal, seen in front of a Hunter fighter-bomber, in 1969 or 1970. (via Ali Tobchi)

- 12 MiG-21UM two-seat conversion trainers
- four S-125 Pechora (ASCC/NATO-codename 'SA-3 Goa') surface-to-air missile (SAM) systems
- Zlin Z.526F and Z.526FI basic trainers from Czechoslovakia
- 10 refurbished MiG-15UTI jet trainers from the USSR[1]

Deliveries of additional MiG-21s enabled the IrAF to significantly bolster existing interceptor units, including Nos. 9, 11, and 14 Squadrons, and to expand No. 17 Squadron that served as the operational conversion unit (OCU) for this type. The fleet of Su-7s was expanded as well; originally, the type was operated by Nos. 1 and 5 Squadrons, both of which were based at Firnas AB, but frequently operated detachments at air bases elsewhere around the country. However, a third unit came into being after the signing of the Friendship Treaty with the USSR: No. 8 Squadron – formerly flying Ilyushin Il-28 bombers – was re-equipped with Su-7BMKs and henceforth served as an OCU for the type. In the place of this squadron, an entirely new multi-unit bomber element came into being – colloquially known as the 'Strategic Brigade' (also 'Tammuz

Table 2: IrAF Order of Battle, October 1973			
Unit	Base	Equipment	Remarks
No. 1 Squadron	Hurrya AB (Kirkuk)	Su-7BMK	CO Maj Sami Hussein Alusy; XO Capt Khaldoun Khattab al-Bakr; **deployed to Damascus IAP on 8 October 1973**
No. 2 Squadron	Firnas AB (Mosul)	4 Mi-1, 31 Mi-4	slated for re-equipment with Mi-8; **involved in the air bridge Syria from 7 October 1973**
No. 3 Squadron	Muthanna AB (Baghdad)	2 de Havilland Dove, 2 de Havilland Heron, 2 Tu-124	VIP transport
No. 4 Squadron	Hurrya AB	9 Westland Wessex HC.Mk 52, 5 SE.316B Alouette III	CO Col Hassan Sharif; slated for re-equipment with Mi-8
No. 5 Squadron	Hurrya AB	Su-7BMK	CO Maj Salim Sultan Abdullah; XO Maj Hassan Kasim Hazem; **deployed to Bley AB on 8 October 1973**
No. 6 Squadron	Taqaddum AB	12 Hunter F.Mk 59	CO Maj Youssef Muhammad Rasooli; **deployed to Qwaysina AB as part of No. 66 Squadron EAF**
No. 7 Squadron	Hurrya AB	MiG-17F	CO Lt Col Shehab al-Qiyasy, XO Capt Mohammed Omar Abdul Hadi; **deployed to Dmeyr AB on 8 October 1973**
No. 8 Squadron	Firnas AB	Su-7BMK	CO Capt Jawdat an-Naqeeb; unit converted to Su-7 in mid-1973 and acted as OCU for that type; **deployed to Dmeyr AB on 9 October 1973**
No. 9 Squadron	Wallid AB	MiG-21PFM	CO Maj Namik Muhammad Saadallah; **deployed to Dmeyr and Tsaykal ABs on 8 October 1973**
No. 10 Squadron	Taqaddum AB	8 Tu-16	
No. 11 Squadron	Rashid AB (Baghdad)	MiG-21MF	CO Lt Col Mohammad Salman Hamad al-Faraji; remaining MiG-21F-13s stored; **deployed to Nassiriya AB on 12 October 1973**
No. 14 Squadron	Rashid AB then Ali Ibn Abu Talib	MiG-21PFM	slated for re-equipment with MiG-21MF
No. 17 Squadron (OCU)	Rashid AB	MiG-21FL, MiG-21UM	CO Col Durhit Ibrahim
No. 18 Squadron	Taqaddum AB	6 Tu-22, 1 Tu-22U	CO Lt Col Mahdi Muhsin asSabbagh; aircraft delivered in early October 1973
No. 23 Squadron	Rashid AB	8 An-12, 12 An-2, 3 Bristol Freighter T.Mk 170	involved in air bridge to Syria from 7 Oct
No. 24 Squadron	Rashid AB	16 Mi-6	CO Maj Shihab al-Jabouri; working up
No. 29 Squadron	Taqaddum AB	12 Hunter F.Mk 59	CO Col Mohammed Jassam al-Jabouri; slated for re-equipment with MiG-23BN; **deployed to Qwaysina AB as part of No. 66 Squadron EAF**
No. 36 Squadron	Taqaddum AB	6 Tu-22, 1 Tu-22U	CO Col. Esmat Shawkat al-Khafar; working up on new aircraft
No. 70 Squadron	Rashid AB	MiG-21M/MF	working up on new aircraft
SAR Flight	Firnas AB	2 Mi-4	slated for re-equipment with Mi-8
SAR Flight	Shoibiyah AB (Basra)	2 Mi-4	slated for re-equipment with Mi-8
Air Force Academy	Sahra AB (Tikrit)	Zlin Z-526 (basic training), L-29 (basic jet training), MiG-15UTI (advanced jet training)	Last Hunting Percival Provosts retired from service in 1971
Flight Leaders School	Rashid AB	6 Hunter T.Mk 66, 17 Jet Provost T.Mk 52	Including Weapons and Tactics School
Bold print denotes a unit's participation in the October 1973 war.			

Wing') – consisting of Nos. 18 and 36 Squadrons, both of which started their conversion to the Tupolev Tu-22. Finally, by 1973 the IrAF began receiving its first Mil Mi-6 and Mi-8 helicopters from the USSR, which were planned to replace old Mi-4s.[2]

Eventually, all of these acquisitions surpassed the capabilities of the Air Force Academy – despite the fact that it was not only re-equipped with Zlins and Delfins, but also redeployed to the newly constructed Sahra AB, outside Tikrit, in 1970–1974. Indeed,

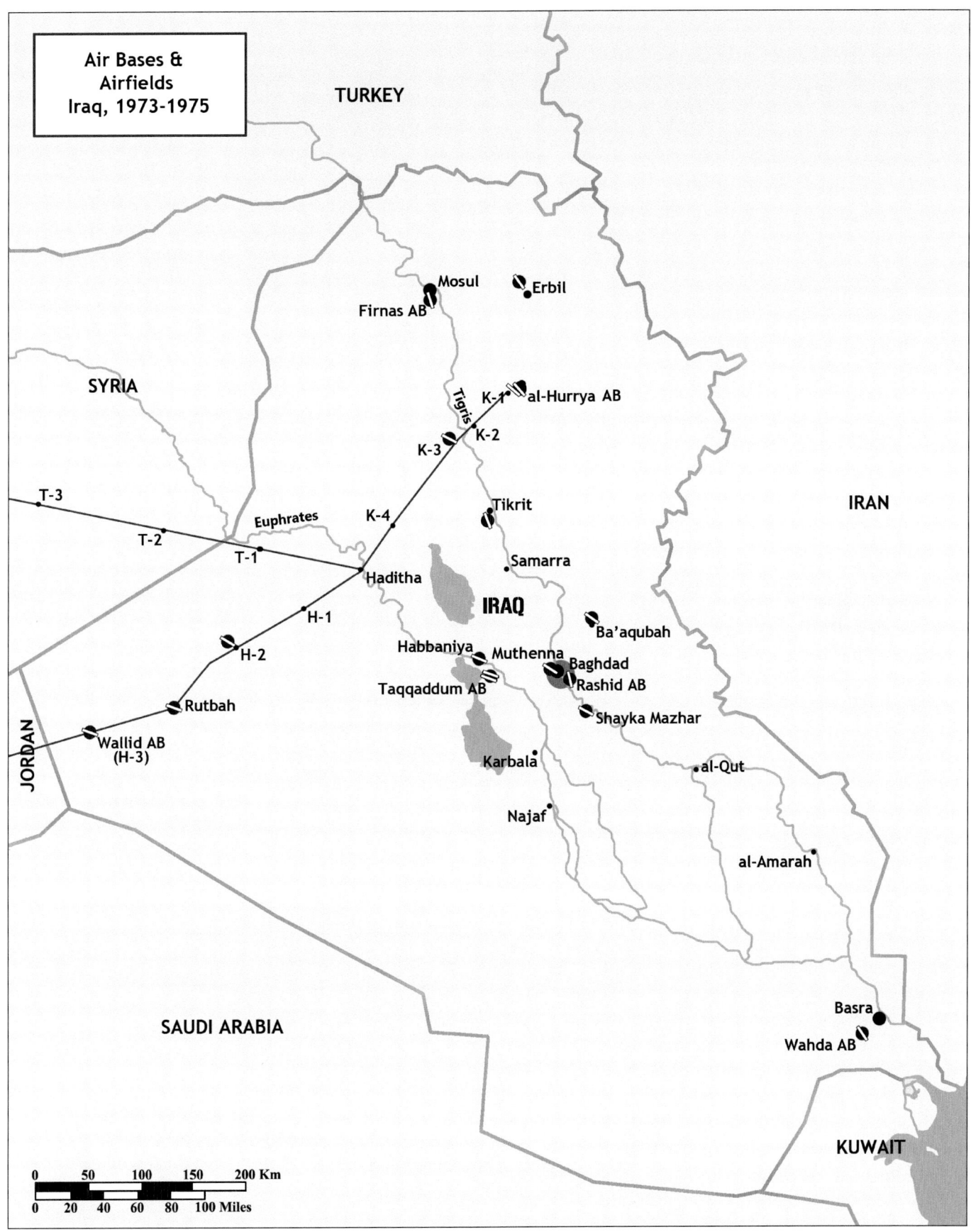

although Baghdad not only invested heavily in having dozens of its pilots trained to serve as instructors, but also expanded the (hitherto 'traditional') presence of pilot instructors seconded from the Indian Air Force (IAF), the Air Force Academy proved unable to train enough new fliers, or to help convert enough of them to fly the newly-purchased combat aircraft. Even the decision to establish a fighter-weapons school – initially equipped with MiG-15UTIs and British-made Hunting Jet Provost T.Mk 52s, and officially designated the Flight Leaders School (FLS) – was considered insufficient for the demand. Correspondingly, additional and ever-larger groups of Iraqi students were sent to Czechoslovakia and the USSR for training purposes. Nevertheless, the construction of Sahra AB, its extensive educational facilities and – later during the 1970s – several 'satellite' airfields nearby, was to have positive consequences for the

A pair of MiG-21MFs (the nearest jet carried serial number 1095) seen from an airliner they were escorting, in the early 1970s. (Ahmad Sadik Collection)

Iraqi pilots undergoing conversion training to the MiG-15bis in Czechoslovakia in the late 1960s, posing with a Czechoslovak Air Force MiG. Eigth from the left in the rear row is Noubar Abdel-Hamid al-Hamadani: the first IrAF pilot officially killed in action against Iran in September 1980.(via Ali Tobchi)

Four IrAF pilots with their MiG-15bis: old high-altitude interceptors, refurbished and reconditioned to serve as advanced trainers and fighter-bombers before delivery to Iraq in 1971. (Albert Grandolini Collection)

of future IrAF pilots, but would also prove its mettle in combat against Kurdish insurgents.³

The same was valid for the decision to expand the H-3 airfield in western Iraq into a full base: by 1970, this had extended runways, underground command and storage facilities, and hardened aircraft shelters, and thus became capable of supporting a much larger number of aircraft which were also heavier than those before. This installation was now not only a hub for a possible deployment of the IrAF to Jordan or Syria in the event of another war with Israel, but became an official air base, named al-Wallid, at which at least a squadron of interceptors was permanently based.

This additional equipment changed the overall capabilities of the IrAF very little in the early 1970s. The primary reason was not only its relatively low status within the Iraqi armed forces' hierarchy: the position of the Chief of Staff in Baghdad remained firmly under control of the army, and most of his closest aides were from the branch specialising in ground warfare, with little knowledge of modern air power. At least as important was the fact that Brigadier General Hussein Hiyawi Hamash at-Tikriti lacked the qualifications for his job, and thus the necessary vision – which was a fact ultimately even Bakr and Saddam were forced to accept. Correspondingly, in June 1973 he was replaced by Major General Nimma Abdullah ad-Dulaymi, a senior pilot who had flown Venoms and Hunters in the late 1950s, before converting to MiG-15s, MiG-17s, and MiG-21s, and in the 1960s becoming a strong proponent of Soviet-made fighter designs. Much to his luck, Dulaymi then decided to retain Hiyawi's staff, foremost among these included the highly experienced Colonel Hamid Sha'ban at-Tikriti, Deputy Operations, who distinguished himself during numerous crises of the previous decade. Almost by accident, the IrAF thus avoided fighting two

IrAF in the long term with the decision to place an order for 50 Aero L-39C/ZO Albatross training jets from Czechoslovakia in 1972. Although the deliveries of these did not occur until 1975 and 1976 (because Aero Vodchody was busy manufacturing L-39s for the USSR), this easy-to-maintain type proved highly popular and was to become not only the basic tool for training nearly 20 generations

Officers of No. 8 Squadron at Firnas AB in 1964, when the unit was commanded by Major Jafar Kamar. The squadron converted to Su-7BMKs in 1973, while the last 6-7 Il-28s were withdrawn from service in 1975. (via Ali Tobchi)

FROM HABBANIYA TO QWAYSINA …

During the 12th Session of the Arab Collective Defence Council, held in November 1971, Chief of Staff of the Egyptian Armed Forces, Lieutenant General Saad el-Din Mohamed el-Husseiny el-Shazly and the top brass of other armed forces of the Arab League had reached an agreement that selected military units from all the member-states were to deploy to Egypt and Syria in the event of another war with Israel. Correspondingly, in early 1973, the General Command in Cairo invited the IrAF to deploy two of its fighter-bomber units to Egypt – supposedly for joint exercises. The selection fell upon the only two units of the Iraqi Air Force that were 'free' of any operational obligations at the time: Nos. 6 and 29 Squadrons, both equipped with Hawker Hunter F.Mk 59A/Bs and home-based at Habbaniya AB. On 18 March 1973, 35 pilots from the two units – many fresh from advanced training courses in the UK, and all with experience from fighting the Kurds – and more than half of the ground personnel were ordered into the main conference hall of Habbaniya AB. Pilots were then handed navigational maps covering northern Saudi Arabia all the way to the Red Sea and eastern Egypt, and briefed about the coming mission. On 6 April 1973, three formations of eight Hunters, and one of three Hunters (the latter serving as spares), accompanied by four

major wars with an unimaginative Air Chief, and without a qualified Director Operations.

Lieutenant General Saad el-Din Mohammed el-Husseiny el-Shazly, Chief of Staff Egyptian armed forces (left), Major General Shanshal, and (shaking hands with Shazly), with Brigadier General Hussein Hayawi Hamash at-Tikrity. By the time of their meeting, it was already certain the IrAF would send two squadrons of fighter-bombers to Egypt to fight the next war against Israel. (via Ali Tobchi)

A Hunter F.Mk 59 (serial 660) of the IrAF, seen in the early 1970s. Two squadrons equipped with this highly popular type were deployed to Egypt in April 1973. (David Nicolle Collection)

Antonov An-12 transports, took off from Habbaniya for the first leg of their trip west – to Tabuq, in Saudi Arabia. The next morning, all four formations then flew to Numan Island in the Red Sea. After refuelling, they continued their trip to Luxor in Egypt. Finally, on 8 April 1973, after flying north along the Nile, and passing the pyramids of Giza, the four formations arrived at Qwaysina AB – a relatively small installation and still partially under construction – in the Nile Delta. After enjoying a day of rest and spending two days with inspections and maintenance of their aircraft, on 11 April all the pilots were gathered in one of the squadron ready rooms for a conference with Major General Hosni Mubarak, Chief of Staff of the Egyptian Air Force (EAF). Mubarak explained to the Iraqis that for the duration of their six-month-long stay in Egypt, their two units would be amalgamated and officially designated No. 66 Independent Squadron, EAF. Over the following weeks, the Iraqis were subjected to a constantly intensifying training syllabus, which first pitted them in mock air combats against one, then two, and finally four Egyptian MiG-21s, and then saw them training in navigation and flying air strikes against simulated targets at ever-lower altitudes. On 28 April 1973, the two IrAF units at Qwaysina were reinforced with four additional Hunters, two of which were two-seaters, and then the decision was taken to bring the families of all the Iraqis deployed in the country to Egypt.[4]

… AND FROM KIRKUK TO DAMASCUS

In late spring 1973, the GMID 'got wind' about the Egyptians and Syrians 'preparing something'. Correspondingly, the Ministry of Defence in Baghdad issued an order for the IrAF to prepare a contingency plan for a deployment to Syria. Nothing happened in this regard until Hamash was replaced by Dulaymi in June 1973: the Iraqis knew no details, because neither Cairo nor Damascus would inform Baghdad. Moreover, the United Arab Command – a body established in 1964 with the aim of auditing and coordinating all the Arab armies in respect of their organisation, equipment and planning for the conflict with Israel – had survived the June 1967 Arab-Israeli War in the form of the Joint Eastern Command which continued to operate for three years longer, but it was disbanded in October 1970 over disagreements between Iraq, Jordan and Syria. Ever since, the three armed forces undertook no joint contingency planning whatsoever. Nevertheless, one of the first things Dulaymi did after taking over the command of the IrAF was to order his staff to prepare plans for the redeployment of five squadrons of fighter-bombers to Syria on short notice.[5]

It thus came as quite a surprise for the Iraqis to find out – mostly from radio news – that a new war between Egypt, Syria, and Israel had erupted, in the afternoon of 6 October 1973. Despite Dulaymi's contingency plan, the outbreak of that conflict could not have caught the IrAF at a worse moment: large elements of the force were still midway through conversion to new aircraft, many pilots and top officers were scattered between conversion and combat courses in Czechoslovakia, Iraq, and the Soviet Union or were undergoing staff courses. Deliveries of MiG-21M/MFs were only just starting seriously, and the majority of the interceptor force was still flying older MiG-21 variants. MiG-17s – delivered back in 1958–1959 – were still in service with No. 7 Squadron, their crews impatiently expecting conversion to MiG-23BN fighter-bombers; and none of the much-expected MiG-23MS were delivered by that date. Indeed, their crews were about to start their conversion courses in the USSR. The only new aircraft ordered in 1972 already in Iraq were 12 Tu-22 bombers and two Tu-22U three-seat conversion trainers, which arrived in early October 1973. Fresh from their training in the USSR, crews of the newly established Nos. 18 and 36 Squadrons were eager about going to war with Israel, and Dulaymi's staff even began planning strike sorties on Israeli air bases. However, all such plans were scratched by the Soviets, who refused to deliver the necessary ground equipment and weapons in time. Indeed, these were to arrive only after the ceasefire, dooming brand-new Tu-22s to spend the October 1973 Arab-Israeli War parked at Taqaddum AB. Ironically, they shared their fate with older Tupolev Tu-16 bombers of No. 10 Squadron; the experience of the June 1967 War had shown that sending these via Jordan to attack heavily protected Israeli air bases was practically suicidal. Instead, the first IrAF unit put on alert for deployment into the combat zone was No. 9 Squadron. Halfway through converting to MiG-21MFs, at 1600 hours on 6 October 1973, the unit commander stopped all the training of his crews, and ordered everybody to prepare their MiG-21PFMs for transfer to al-Wallid AB. The first formation took off in that direction at dawn the following morning and by 1640 hours of 7 October, 10 aircraft of the unit were forward deployed at Dmeyr and Tsaykal air bases, in Syria. The squadron flew its first combat air patrols (CAPs) and its pilots had their first contacts with Dassault Mirage IIIs, Mirage 5s and McDonnell Douglas F-4E Phantom IIs barely one hour later.[6]

On 8 October 1973, two Su-7BMK units followed. Aircraft from No. 1 Squadron transferred from Hurrya AB near Kirkuk, via Rashid and al-Wallid AB to Damascus International, while No. 5 Squadron followed along the same route to Bley AB, south of the Syrian capital. Lieutenant Colonel Mohammad Salman Hammad was appointed the commander of the IrAF contingent in Syria, with Major Mohammed Jassim Hanish al-Jaboury as liaison officer for

A pilot and ground crew of No. 5 Squadron in front of the Su-7BMK serial number 769, at Hurrya AB, in Mosul. Notably, the jet was one of few to receive a camouflage pattern before its delivery to Iraq: like all of the Su-7BMKs operated by the IrAF, it had no rear-view mirror atop the cockpit transparency and only one underwing hardpoint per wing. (via Ali Tobchi)

Hardened aircraft shelters at the south-eastern side of Damascus International Airport, which housed Su-7BMKs of No. 1 Squadron, IrAF during the October 1973 Arab-Israeli War. (Photo by Tom Cooper)

rockets, and ammunition for their internal 30mm ADEN cannons; all 16 returned safely to Qwaysina. Within minutes of their return, all 16 aircraft were completely re-armed and refuelled, and their pilots prepared for their next mission. However, the related order never came: the second wave of the opening air strike was cancelled by Mubarak.[8]

Thus, the IrAF Hunters deployed in Egypt flew their second mission only late in the afternoon of 8 October 1973, when eight aircraft attacked an Israeli column east of the Tasa Defile. Flying extremely low, they took the enemy by surprise and found an intersection jammed full of trucks and other 'soft-skin' vehicles: this was exposed to eight attack runs and a total of 192 unguided rockets in quick succession. Numerous explosions were observed and

interceptor operations, and Captain Nagdat Muhammad Mustafa an-Naqeeb as liaison officer for the fighter-bomber units. The contingent was reinforced through the addition of Su-7s from No. 8 Squadron and MiG-17s from No. 7 Squadron, on 9 and 10 October, respectively. During the transfer – supported by An-12 and An-24 transports – all the units were reinforced by officers reassigned from the Staff College and other training institutions, thus reaching a ratio of 1.5 pilots for every aircraft.[7]

ACTION OVER THE SINAI

By the time the redeployment of IrAF units to Syria began, the first Iraqis had already seen combat against Israel. Indeed, the first mission involving IrAF aircraft was flown as a part of the opening Egyptian air strike against enemy positions on the Sinai Peninsula, at around 1400 hours on 6 October 1973; it included 12 Hunters of No. 66 Squadron that attacked a MIM-23 HAWK SAM site near the Gidi Defile, and four that rocketed the camp of an Israeli artillery unit nearby. All jets were armed with a full load of 24 unguided 127mm

Captain Wallid Abdul Latif as-Samarrai, Deputy Commander No. 66 Squadron, one of two Hunter pilots killed during a mission flown in support of the Second Field Army, late in the afternoon of 8 October 1973. (Ali Tobchi Collection)

1st Lieutenant Amer Ahmad Qaysee was the pilot of the second IrAF Hunter shot down during the late afternoon of 8 October 1973. (Ali Tobchi Collection)

A pair of Iraqi Hunters seen during the October 1973 Arab-Israeli War streaking very low over the high earth berms constructed by the Egyptians and Israelis along the Suez Canal. (Tom Cooper Collection)

the target area left covered with smoke. On the negative side, the Hunter flown by Captain Wallid Abdul Latif as-Samarrai, Deputy Commander No. 66 Squadron, was shot down and he was killed. Another jet, piloted by 1st Lieutenant Amer Ahmad Qaysee, was then hit by Egyptian air defences while the formation was returning to the western side of the Suez Canal. The young pilot was never recovered.[9]

QUNAITRA OR BUST

The situation Iraqi pilots found in Egypt and Syria was quite unusual for them. In addition to differences of an organisational nature, and when it came to the activity of the Israeli MIM-23 SAM sites on Sinai, both hosts lacked precise targeting intelligence of the enemy, and reconnaissance photographs of most objectives were not available. The Iraqis were used to operating with the support of forward air controllers – officers equipped with a radio and either deployed with ground troops or flying light aircraft like Cessna O-1 Birddogs – and the EAF is known to have used similar tactics during the mid-1960s in Yemen, yet in 1973 this concept was unheard of in both Egypt and Syria. Instead, both the EAF and the SyAAF flew under orders issued by general commands in Cairo and Damascus. Most of these reached the respective units with such a delay that by the time the aircraft reached the area in question, the target was long since gone. Other differences were related to the equipment of their aircraft. For example, the Iraqis flew their Su-7BMKs still in their original configuration, with no rear-view mirrors atop the cockpit transparency, and with only one hardpoint per wing. Along with modifications originally introduced in Egypt in the 1967–1968 period, Egyptian and Syrian Su-7BMKs were meanwhile re-equipped with one additional hardpoint per wing (for a total of six, including the two under the centre fuselage), and had a rear-view mirror atop the cockpit transparency, greatly increasing the pilot's field of view. Finally, later during the war, the Iraqis noted the appearance of brand-new, hitherto unknown Soviet-made air-to-air missiles in Syria; the weapons in question were probably early examples of the R-13M (ASCC/NATO-codename 'AA-2c Advanced Atoll'), which was to reach Iraq only years later.

That said, the Iraqis found their Syrian hosts particularly determined, ready to sacrifice virtually everything to recover the Golan Heights. Combined with their indoctrinated pan-Arabism, this bolstered the morale of both IrAF and SyAAF fliers and was to explain much of their behaviour over the following two weeks. For example, MiG-21 pilots of No. 9 Squadron sought every opportunity to engage Israeli fighter-bombers underway over Syria. After studying their failures from the late afternoon of 8 October 1973, they quickly adapted and on the next morning Lieutenant Colonel Mohammed Salman Hamad al-Faraj and 1st Lieutenant Kamil Sultan al-Khafaji were directed by the ground control into an attack on four Mirages flying a CAP over the Golan Heights. While Hadid managed to sneak unobserved upon the enemy and hit one with

The pilot of a Su-7BMK, seen on return to Iraq in November 1973. The camouflage colours were applied to the aircraft in Syria during the October 1973 Arab-Israeli War. (Ahmad Sadik Collection)

two R-3S air-to-air missiles (ASCC/NATO-codename 'AA-2 Atoll'), Khafaji became isolated, was involved into a turning engagement with multiple opponents, shot down and killed.[10]

WASTED OPPORTUNITIES

Under the given circumstances, Lieutenant Colonel Faraj was initially overcautious with his orders for combat operations. Aiming to gradually accustom his pilots to the battlefield and the combat zone, early on 9 October he dispatched a pair of Su-7BMKs from both Bley and Damascus on an 'armed reconnaissance'; their pilots had no specific targets, but were to cross the Golan Heights and then fly down the Jordan River and attack targets of opportunity. Armed with four FAB-250M-62 bombs each, the jets passed the frontline unmolested and quickly reached the Jordan River, only to aimlessly continue down its flow. The pair, piloted by Captain Khaldoun Khattab al-Bakir and 1st Lieutenant Noubar Abdul, thus dashed low over the B'not Ya'acov Bridge, entirely undisturbed, but did not attack. Similarly, the pair piloted by Captain Muwaffak Saeed Abdullah an-Naimi and 1st Lieutenant Mohammed Mohsen as-Saydoon reached the Arik Bridge and then the B'not Ya'acov Bridge without problems. All four pilots reported lots of Israeli traffic on the ground, but they missed the golden opportunity; had they been ordered to attack one of the bridges, they could have caused a major delay in the flow of IDF (Israeli Defense Force) reinforcements to the Golan Heights, thus buying time for forward Syrian Army units to better secure the areas they had brought under control. Instead, all four dropped their bombs upon a column of Israeli tanks near el-Al, with unknown effects. Moreover, while crossing back into the Syrian-controlled territory, all four jets came under friendly fire, and both Bakir's and Hamid's Sukhois received multiple hits.[11]

Emboldened by the lack of Israeli resistance, Hammad ordered the first combat sortie; a few hours later, four Su-7BMKs – including the example piloted by Captain Alwan al-Aboossi – launched from Dmeyr AB. They were led by the Syrian Colonel Abdul Rahman in a fifth Su-7BMK, who knew the combat zone very well, and quickly found a column of Israeli tanks and APCs to unleash their FAB-500M-62 bombs upon. Longing for action, Captain Khaldoun Khattab al-Bakir then led three wingmen into an attack on the

Captain Salam Mohammed Ayoub was the second IrAF Su-7 pilot shot down during the October 1973 War. He ejected safely but was murdered by soldiers of the Moroccan Expeditionary Brigade. (via Ali Tobchi)

Israeli headquarters in Kfar Naffech. The Number 3 of this flight, 1st Lieutenant Mohammad Alwan, disappeared without a trace while underway to the target, and was never seen again. The other three aircraft reached their target and bombed it with a total of 12 FAB-250M-62 bombs.[12]

The last known mission of the day by IrAF Su-7BMKs resulted in yet another loss. When a formation of these was – finally – sent to bomb the B'not Ya'acov Bridge, the jet piloted by Captain Salam Mohammed Ayoub was hit by ground fire. Although the pilot managed to nurse the stricken aircraft back over Syrian frontlines, and ejected safely, he parachuted into the positions of the Moroccan Expeditionary Brigade deployed near Jebel Sheikh. Notorious for their savagery, the Moroccans misidentified the young Iraqi as 'enemy' and murdered him.[13]

FRIENDLY FIRE PROBLEM

After spending 9 October 1973 mourning the loss of two pilots and studying the reasons and attempting to improve coordination with the air defence system of the Egyptian Army, No. 66 Squadron was about to go into action again early on 10 October. Just at that moment in time, the IDF/AF hit Qwaysina AB with a formation each of F-4E Phantom IIs and McDonnell Douglas A-4 Skyhawks. The incoming attack was detected in time, and all the Iraqis were well protected within underground shelters. Listening to the sound of dozens of detonating bombs, they expected to find their air base completely demolished. However, once back on the surface, everybody concluded that except for a few windows at the squadron ready room and the weather station, the damage on the well-fortified base was minimal. One of the ground crew was slightly wounded, and one of the Hunters – all of which were parked inside hardened aircraft shelters – lightly damaged. Quickly-applied repairs returned not only the jet, but the entire base back to operational status in a matter of minutes. Nevertheless, orders for further action arrived only early on 11 October, when a formation of eight Hunters led by Major Mohammed Ali was ordered to attack a concentration of Israeli troops opposite the Third Field Army in two waves. Much to the disgust of everybody within No. 66 Squadron, once again, the mission was spoiled by overzealous anti-aircraft crews of the Egyptian Army. Major Ali's Hunter was badly damaged while approaching the Suez Canal Zone; he managed to nurse it back for an emergency landing at Qwaysina. However, the jet piloted by 1st Lieutenant Diah was shot down. The pilot ejected safely, but was then badly beaten by Egyptian Army troops because of his British-made flying suit, and ended up in a hospital. Certainly enough, the remaining two Hunters pressed their attack home, and hit one of the ordnance companies of the Israeli army, destroying several fuel trucks and causing a conflagration that demolished additional vehicles. However, in the light of negative experiences with Egyptian anti-aircraft defences, the second quartet of Hunters was not even launched, pending further investigations.

ALL-OUT EFFORT

On 10 October 1973, the two Iraqi Su-7 units in Syria flew more than 30 combat sorties, bombing Israeli ground forces all the way from B'not Ya'acov Bridge to Qunaitra. On the contrary, in the light of continuous Israeli air strikes on Syrian air bases, and heavy losses of SyAAF interceptors, Iraqi MiG-21s were meanwhile held back and scrambled only when an engagement was certain. Unsurprisingly, the pilots of No. 9 Squadron are known to have had only one engagement with the fighter-bombers of the IDF/AF that day: Captain Ali Hussain and 1st Lieutenant Fayez Bagir from No. 9 Squadron had intercepted a pair of F-4E Phantoms but found themselves exposed to attacks by multiple air-to-air missiles and were forced to disengage. The Sukhoi pilots of No. 1 Squadron were less lucky. They lost the jet flown by 1st Lieutenant Ahmed Saleh Muhammad al-Obeidi, which was shot down by Israeli ground-based air defences and seen to impact the ground with the pilot still inside the cockpit.[14]

By 11 October, Syrian ground forces were pushed away from the Golan Heights and forced to withdraw from Qunaitra in the direction of Damascus. Too proud to admit another setback, the General Command in Damascus ceased issuing clear orders, instead demanding Iraqi and allied fliers to, literally, 'attack whatever enemy forces they could find'. The Syrian generals lost control of the battle, and their failure to inform everybody involved had a very negative impact upon the morale of pilots and ground crews. The Iraqis continued fighting, nevertheless, and their Su-7 units had another 'all-out day', conducting about 30 combat sorties, and suffering one recorded loss: the jet piloted by 1st Lieutenant Reda'a Jamil Hussain – one of fliers re-qualified to operate Sukhois in Czechoslovakia – was shot down by enemy ground defences and killed.[15]

Still, and although exhausted by days of intensive flying and fighting, the Iraqis – as well as their Syrian- and Egyptian comrades deployed in Syria – continued flying and fighting, although lacking clear information about where the friendly and enemy forces were. On 12 October 1973, No. 7 Squadron, IrAF suffered its sole combat loss of the war, when a flight of MiG-17Fs was caught by Israeli Mirages while attacking enemy positions in the Qunaitra area, and the jet piloted by 1st Lieutenant Issam Ali Jinkir was shot down,

1st Lieutenant Ahmed Saleh Muhammad al-Obeidi, a Su-7BMK pilot from No. 1 Squadron, shot down and killed over the Golan Heights on 10 October 1973. (via Ali Tobchi)

1st Lieutenant Reda'a Jamil Hussain, a Su-7BMK pilot from No. 1 or No. 5 Squadron, shot down and killed over the Golan Heights on 11 October 1973. (via Ali Tobchi)

A MiG-17F serial number 724 of the Iraqi Air Force, seen upon return to Iraq. The camouflage pattern, which had been applied in Syria in a great rush, was already badly worn out by heavy use. (via Ali Tobchi)

Found amid the rubble of Qunaitra after October 1973, the fin of this MiG-17 obviously shows the Iraqi flag, and traces of the serial ending with 27/37 or 67 (visible near the tip of the fin, on the left side of the photograph). Because No. 7 Squadron IrAF lost only one jet in combat, this was probably the MiG-17F piloted by Jinkir when he was shot down, on 12 October 1973. (Photo by David Nicolle)

1st Lieutenant Issam Ali Jinkir was the only pilot of No. 7 Squadron IrAF killed in action during the October 1973 War with Israel. (Ali Tobchi Collection)

forcing the pilot to eject. Led by Major Namiq Sa'adallah, the escorting MiG-21s were slow to react, and although Sa'adallah claimed a Mirage shot down by an R-3S missile, Jinkir was never found again. Later during the day, the IrAF contingent in Syria was reinforced by the arrival of 12 brand-new MiG-21MFs from No. 11 Squadron; these were redeployed to an-Nasiriya AB.[16]

The Israelis continued striking air bases in south-western Syria throughout that time, constantly increasing pressure; on 13 October 1973, F-4Es bombed Bley AB, where their bombs completely demolished one of the hardened aircraft shelters with an Iraqi Su-7BMK inside. Furthermore, a brand-new MiG-21MF from No. 11 Squadron was shot down in air combat with Israeli interceptors, together with its pilot, 1st Lieutenant Mutayb Ali az-Zobai. In turn, the Syrian ground control managed to direct a pair of MiG-21PFMs from No. 9 Squadron to intercept Israeli A-4 Skyhawks operating along the frontlines, and one of the Iraqi pilots claimed a fighter-bomber shot down by two R-3S missiles. Later during the day, a pair of MiGs from the same unit led by Major Ma'an Abd ar-Razzaq al-Awsi found themselves in a strange situation. After being directed towards an incoming Israeli formation, as soon as they reported having the enemy in sight, both pilots were ordered to fly a 180-degree turn and make distance at maximum speed. As they darted back in the direction of Damascus, they saw three white contrails streaking past them and upwards; later on, the Syrians explained to them that their ground control used the two MiGs as bait for a formation of Israeli Mirages, one of which was shot down and its pilot captured.[17]

That said, the combination of sustained air strikes and bitter resistance by Syrian Army units on the ground eventually bought time for the 3rd Armoured Division of the Iraqi Army to reach the frontlines – just in time to launch a counterattack into the right flank of the IDF's advance on Damascus. Although suffering losses in a series of pitched battles fought through 12 and 13 October, the Iraqi troops forced the Israelis to stop and dig in.

Pilots of No. 9 Squadron with their CO, Major Namiq Sa'adallah (centre, with sunglasses), in front of one of that unit's MiG-21PFMs (serial 840) wearing camouflage colours applied in Syria, seen after their return to Iraq in 1974. (via Ali Tobchi)

Pilots of No. 11 Squadron, IrAF, in December 1968, in front of one of their MiG-21FLs: by 1973, the unit was re-equipped with MiG-21M/MFs. (via Ali Tobchi)

Major Ma'an Abd ar-Razzaq al-Awsi, one of two Iraqi MiG-21 pilots used on 13 October 1973 as bait by the Syrian air defences to drag Israeli Mirages into a SAM-trap and shoot one of them down. Considered a very fine pilot, Awsi continued a successful career with the IrAF and later served with MiG-23MS-equipped No. 39 Squadron until shot down and killed early during the war with Iran in December 1980. (IraqiAirForceMemorial.com)

DAY OF THE HUNTERS

Meanwhile in Egypt, it took two days for the staff of No. 66 Squadron attempting to improve coordination with the Egyptians, before No. 66 Squadron received a new air tasking order. On 13 October 1973, it received an order to attack Israeli tanks deployed on the eastern side of the Suez Canal, opposite Deversoir. Two flights of four jets were launched, one led by Captain Emad Ahmed Ezzat, and the other by Captain Mohammed Naji. This time, Hunters passed over the Egyptian Army positions without problem, and then followed a road in the direction of their target. Upon reaching the crossroad that they were to attack, the Iraqis found no targets, and made a 90-degree turn north. Seconds later, they saw a large convoy of vehicles and promptly attacked from convergent directions. Their 192 rockets caused tremendous detonations and a massive conflagration, the shockwaves of which caused all the jets to shudder, but at first it appeared the Iraqis suffered no losses. Only once they were on the way back did Captain Ezzat report that his aircraft was hit: he ejected and was subsequently captured by the Israelis. Back in Qwaysina, the survivors then gathered to debrief and try to figure out what had happened. It transpired that the Hunters had hit a column of trucks hauling artillery ammunition as it was passing the positions of the 329th Reserve Artillery Battalion of the IDF, equipped with US-made M107 self-propelled howitzers and Czechoslovak-made OT-62 armoured personnel carriers (captured from Egypt in 1967). One of the jets was either shot down or caught in the conflagration, and crashed into the Israeli position, further increasing the damage caused by unguided rockets. While its pilot, 1st Lieutenant Sami Fadil (Ezzat's Number 2) was killed, the IDF unit suffered such losses in men and material that an entire company of the 329th Battalion was disbanded. As far as is known, this remains the most massive single blow the Iraqi armed forces have ever delivered upon Israel or its armed forces.[18]

The Hunters of No. 66 Squadron, EAF, were in action again a day later, when sent to attack an Israeli unit east of el-Qantara. Very few details about this operation are known, however, except that the formation lost two jets and their pilots. The reason for the loss of Captain Abdul Qadir Haydar's Hunter remains unknown; reportedly, he ejected safely but was killed by Israeli troops on the

1st Lieutenant Sami Fadil was killed when his Hunter was hit by ground fire and crashed into Israeli artillery positions in the Sinai on 13 October 1973. (via Ali Tobchi)

Hunter pilot Captain Abdul Qadir Haydar, shot down by Israeli ground fire east of Qantara on 14 October 1973. He ejected safely but was subsequently killed by Israeli troops. (via Ali Tobchi)

1st Lieutenant Abdul Qadir Duraid, while serving with No. 6 Squadron at Qwaysina AB in Egypt. (via Ali Tobchi)

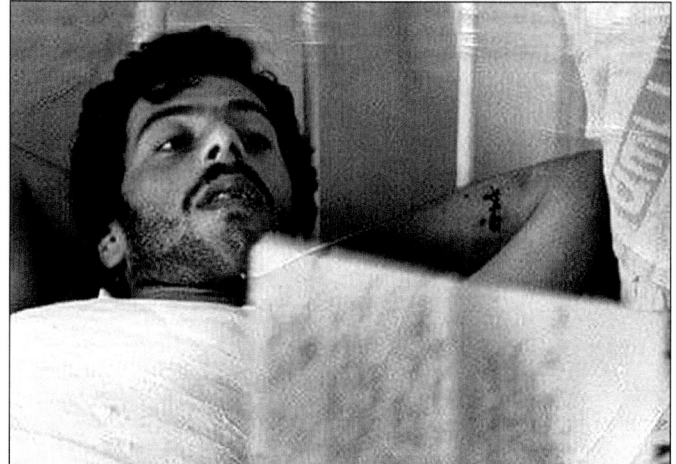

Duraid seen in a hospital in Tel Aviv while a prisoner of war. After the prisoner-of-war-exchange, he returned to serve with the IrAF, and reached the rank of a Brigadier General before retirement. (via Ali Tobchi)

Wreckage of US-made M107 self-propelled howitzers and ammunition trucks of the 329th Reserve Artillery Battalion, IDF, after the Iraqi air strike of 13 October 1973. (Albert Grandolini Collection)

ground. 1st Lieutenant Abdul Qadir Duraid's jet was shot down by anti-aircraft fire: he ejected safely and was taken prisoner of war.[19]

SUDDEN END

The last week of the war on the Syrian front saw a massive surge in Iraqi and Syrian aerial activity; on average, Su-7BMKs of the IrAF flew about 30 sorties a day between 14 and 22 October. The majority of Israeli fighter-bombers were busy on the Egyptian front, but interceptors of the IDF/AF were present in some numbers, especially during the final battle for the two peaks of Jebel Sheikh. The Israelis credited their pilots with more than 40 aerial victories on 21, 22 and 23 October in this area alone; however, next to nothing is known about specific operations of the IrAF, except that Major Namiq Sa'adallah claimed another Mirage as shot down in air combat on the 23rd. A day later, Damascus accepted the UN-negotiated ceasefire, prompting a fierce reaction from Baghdad: already bitter about Egyptian and Syrian refusals to inform Iraq about their attack, and then the Egyptian acceptance of the ceasefire, and knowing this would enable Israel to concentrate all of its power upon Syria, Bakr and Saddam reacted by ordering an immediate withdrawal of all of their units from Syria – including the SyAAF contingent. The last Iraqi airmen were all back home by 30 October 1973.

The situation was only slightly different in Egypt, but then for different reasons. After heavy losses of No. 66 Squadron, the General Command in Cairo became reluctant to continue ordering Iraqi Hunters into additional combat. Furthermore, on 17 October 1973, the IDF/AF flew its second air strike on Qwaysina AB, hitting the base of No. 66 Squadron particularly hard. The runway was cratered in numerous places, and even the Deputy Commander of the unit, Major Yusuf Muhammad Rasouli, was badly injured. This time, the Egyptians took hours to restore the base to at least semi-operational condition; in fact, the damage was fully repaired only after the war. Furthermore, IrAF Hunters are not known to have flown any additional combat sorties before the cease-fires of 22, 24, and 25 October 1973.

IRAQI AIR BRIDGE

While Iraqi fighter-bomber pilots earned themselves lots of praise for their achievements in combat, transport units of the IrAF provided a highly valuable, yet usually ignored contribution to the overall war effort. When the October 1973 War erupted, the air force had 31 transport aircraft at its disposal, including four Antonov An-2s, seven An-12s, 10 An-24s and eight Ilyushin Il-14s. Two Tupolev Tu-124s, and two de Havilland Herons were used for VIP transport, but they were pressed into service, too. The theoretical total cargo-lift capacity of these aircraft was only 145 tons – or 728 fully-armed troops, or, with some 'field modifications', up to 1,584 soldiers. Obviously, this was a meagre capability in comparison with other involved air forces, especially considering the slow speed of most of the aircraft in question; heavily loaded An-2s took at least 17 hours for just one trip from Baghdad to Damascus.[20]

Nevertheless, all routine operations were cancelled and all aircraft and all available crews pushed to fly as often as possible, many loaded beyond their safety limitations. Overall, during the October 1973 Arab-Israeli War, Iraqi transports undertook 862 sorties, of which 121 were in support of IrAF units in Egypt and 394 in Syria, while 219 sorties were flown in support of Egyptian, Syrian and Iraqi Army units. They transported 1,800 personnel, 671 tonnes of cargo to Egypt and 452 to Syria. Moreover, even Mil Mi-4 helicopters were pressed into operations related to the redeployment of Iraqi Army units underway from Baghdad to Damascus and flew 83 related sorties.[21] Despite this all-out effort, the IrAF subsequently found itself exposed to fierce critique. Some blamed it for failing to use its helicopters for the deployment of special forces beyond enemy frontlines, and others for the failure to support a faster redeployment of the Army's 5th Mountain Brigade into a counterattack on Jebel Sheikh, on 20–22 October. Actually, neither was possible because all available aircraft and helicopters were already preoccupied with supporting other operations. Nevertheless, such experiences strongly influenced the subsequent development of the transport branch of the air force.

STATISTICS AND LESSONS LEARNED

During the October 1973 Arab-Israeli War, the Iraqi armed forces suffered a loss of 278 officers and other ranks killed in action, and 898 wounded in action. While flying a total of 862 hours in combat – of which 121 over the Egyptian front and 394 over the Syrian front – the IrAF lost 12 pilots and four were taken as prisoners of war: Major Hussein Rashid from No. 7 Squadron, Captain Imad Ahmed Ezzat from No. 66 Squadron, 1st Lieutenant Abdul Qadir Duraid, and 1st Lieutenant Sa'ad Abd al-Muhsin al-Azmi from No.

Seven Iraqi An-12s from No. 23 Squadron formed the backbone of the IrAF transport fleet and were intensively involved in the October 1973 War. They mostly hauled personnel and spares from home-bases in Iraq to Egypt and Syria. (Albert Grandolini Collection)

All 31 Mi-4s of No. 2 Squadron, IrAF, were deployed in support of the Iraqi Army's deployment to Syria. Among others, they hauled teams of Army technicians that provided technical support for troops of the 3rd Armoured Division, marching along the highway from Baghdad via Ruthba to Damascus. (Tom Cooper Collection)

DUMPING SUKHOIS UPON BAGHDAD

If the crews of the newly established Nos. 18 and 36 Squadrons were bitterly disappointed about the Soviets refusing to deliver the ground equipment and weaponry necessary to operate their 'brand-new' Tu-22s against Israel during the October 1973 War (all of the aircraft in question turned out to have been built in the 1960s, and only refurbished before delivery to Iraq), much of the air force was at least as much astonished by Moscow's next move. On 12 October 1973, midway through the ongoing Arab-Israeli War, 10 An-12 transports of the Soviet Air Force landed at Rashid AB, each loaded with one disassembled Sukhoi Su-20 fighter-bomber. The Iraqis helped the Soviets unload their cargo, loaded the crates on tank transports and brought everything to Hurrya AB, near Kirkuk. The delivery came as a big surprise for everybody involved; while an Iraqi military delegation including several Su-7 pilots had inspected Su-20s at Kubinka AB, a year before, no order was placed. The Soviets literally dumped one of their newest fighter-bombers upon the IrAF. Not to be outdone, the Iraqis did then place an order for additional Su-20s: the resulting fleet of 18 aircraft entered service with Hurrya-based No. 1 Squadron, in early 1974, only months after the unit returned from its deployment in Syria. Before long, the type proved highly popular in service; however, this was to remain the sole acquisition of this early variant by Iraq. In the future, the IrAF would demand – and the Soviets would eventually deliver – much better-equipped sub-variants.[24]

A still from a video showing a section of Su-20s operated by No. 1 Squadron, starting with February–March 1974. Notable are the snow-capped mountains of northern Iraq in the background; they became a part of the unit's new crest, designed around the same time. (Ahmad Sadik Collection)

5 Squadron were all exchanged for Israeli prisoners of war in Syria, and repatriated. The material damage was significant. Thirty combat aircraft were written off between 6 and 24 October, including 15 Su-7BMKs, eight Hunters, five MiG-21s and two MiG-17Fs. On the positive side, three pilots – Mohammad Salman Hammad, Namiq Sa'adallah and Shehab Ahmed – were officially credited with a total of five aerial victories against Israeli fighters (in exchange for four Iraqi jets acknowledged as lost in air combats with the Israelis).[22]

At least in public, the war was a positive experience; not only because the Arabs restored the honour of their armed forces, tarnished during the June 1967 War, but the Iraqi pilots earned themselves much praise, especially in Egypt. As far as is known, every single officer of No. 66 Squadron was highly decorated before leaving Cairo. Less obvious was the fact that the Iraqis drew completely different lessons from the October 1973 War with Israel than Egyptians and Syrians did – and the way in which they did this. Contrary to both the EAF and the SyAAF, the IrAF held a number of post-war conferences during which its officers were encouraged to engage in very direct and frank exchanges of experiences and opinions. Their primary conclusion was that investment into the training of pilots and ground crews, and aircraft acquisition were the best solutions to all air defence related issues. Correspondingly, even though the IrAF subsequently did buy additional SAM systems from the USSR, it was never to become dependent on so-called 'SAM-belts' or 'SAM-networks' for the defence of Iraqi airspace, and would never come to the idea of 'hiding' its manned interceptors behind ground-based air defence arrays. On the contrary, manned fighter-interceptors remained the primary means of air warfare of the IrAF, with ground-based air defences only serving for point defence. Another experience was the necessity to centralise the command and control into a network of sensors and control posts, all input from which would be collected in one place to enable a unified command to control the battle in real time. This was not only necessary to avoid such painful losses from friendly fire as the IrAF experienced over Egypt, but also to better exploit the available firepower, and to better protect its own fighter-bombers from enemy interceptors. The third crucial experience was the Israeli superiority in regards of high technologies and firepower. Not only that US and Israeli-made air-to-air missiles proved superior to Soviet-made missiles – almost rendering the entire MiG-21 fleet ineffective to the degree where it was facing block-obsolescence pending major upgrades – but the Iraqis had never encountered electronic warfare before. This was to deeply influence the further development of their air force over the following decade. Finally, the October 1973 Arab-Israeli War made it clear – not only in Baghdad, but in other Arab capitals, too – that the USA would continue supporting Israel regardless of the consequences for its relations to the Arab world, or even its own economy, and thus fighting further large-scale wars against Israel was in fact pointless. At least in this regard, it can be said that it took the Iraqis a full generation to forget this important lesson.[23]

3
SECOND IRAQI-KURDISH WAR, 1974–1975

In January 1968, amid a major economic crisis in Great Britain, Prime Minister Harold Wilson and his Defence Secretary, Denis Healey, announced a general withdrawal of all British troops from major military bases in southern and eastern Asia, 'east of Suez', by 1971. Even if Britain actually retained a military presence in a few spots – like Oman – this decision had far-reaching consequences long before it was realised. After failing to form the proposed Federation of Arab States, British protectorates on the western coast of the Persian Gulf declared their independence: Bahrain in August, Qatar in September, and the Trucial States on 1 December 1971 as the United Arab Emirates (UAE). These new nations had only small armed forces trained in maintaining public order. Unsurprisingly, they felt vulnerable, and were promptly 'confirmed' as such by Iranian actions. On 29 and 30 November 1971, a task force of the Imperial Iranian Navy landed marines on the three islands in the lower Persian Gulf – Greater and Lesser Tunb, and Abu Mussa – controlled by Iran for millennia before the arrival of the Portuguese in the sixteenth century, but which the British had handed over to the UAE. The tiny Arab police force offered some resistance and killed four of the marines but was subsequently overrun by the superior force. Seeking to establish himself in position of impeccable Arab steadfastness – indeed as new protector, if not the future leader of the Arab world – Saddam Hussein reacted with a fiery speech, announcing an intention to recover the three islands. Simultaneously, the RCC attempted to secure its back by solving the ongoing conflict with the Kurds of northern Iraq; however, while offering autonomy, it declared Iraqi Kurdistan to be an 'integral part of the Arab world' and insisted that Kurdish self-determination would be impossible. When negotiations with Mustafa Barzani failed, the RCC deployed 12 brigades of the army in two offensives aimed to secure the area between Ruwanduz in the north and Sulaymaniyah in the south. Largely successful, these operations forced Barzani to renew talks. In early 1970, Baghdad announced that the Kurds were not only granted de facto autonomy, but also the right to participate in the government. Simultaneously the RCC had a major showdown with leaders of the Shia majority population; their demonstrations of late 1969 were crushed by the army, there was a ban on religious schools and processions, and dozens of thousands of 'illegal Iranian immigrants' were expelled – most of whom turned out to be Kurds – to Iran.

Combined, all of these actions, and especially Saddam's rhetoric, played straight into the hands of Shah Mohammed Reza Pahlavi II of Iran, who was in the process of positioning himself as the closest US ally in this part of the world through announcing he would oppose any 'radical' – and especially 'Soviet-allied' – 'Arab regimes'. For all practical purposes, Iran was now to act as a new 'policeman' of the Persian Gulf. Unsurprisingly considering the centuries-old arch rivalry between the Arabs and Persians, the resulting 'war of words' was soon converted into the next armed conflict involving Iraq.

WAR OF WORDS
Keen to keep Iraq weak, the Shah was clandestinely supporting the rebellion of Mustafa Barzani – at that time a self-appointed leader of the Kurdistan Democratic Party (KDP) – with arms, supplies,

A Hunter F.Mk 59 serial number 664, seen while equipped with the reconnaissance nose with Vinten cameras installed inside (recognisable by the large transparency on the side and the front). Notably, although this photograph was taken in the 1970s on Malta, the nose and the fin were still painted in red – as per the UAC's directive from 1966 (see Volume 1 for details). (Albert Grandolini Collection)

A PHOENIX OF A SQUADRON

Considering the earlier history of MiG-19s in Iraq, and the fact that No. 29 Squadron was quite famous for flying Hunter fighter-bombers since 1964–1965, it was quite surprising to find out that during the summer of 1974, the IrAF reinforced units involved in fighting Barzani's Peshmerga ('those who face death') with a squadron flying MiG-19s.

The reasons for this re-emergence of MiG-19s in Iraq, and for these being operated by No. 29 Squadron, were relatively complex. Probably in relation to the Iraqi-Soviet Treaty of Friendship, in August 1972 Moscow donated a batch of 12 refurbished MiG-19s to Iraq. The aircraft were delivered on board Soviet Air Force An-12 transports to Rashid AB, and assembled by Soviet technicians. All were from the surplus stocks of the Soviet air force; their fuselages were in excellent condition, but their engines were worn out and, much to their dismay, the Iraqis realised that the engines were not interchangeable. There was no way to, for example, replace the left engine on one aircraft, with the right engine from another. Moreover, spares were scarce, because production of MiG-19s in the USSR had already ended in 1959. Eventually, all 12 jets were stored at Rashid AB.

Then, in early 1974, Nos. 6 and 29 Squadrons returned from Egypt and were re-established as independent units. This caused a problem for the Quartermaster General of the IrAF, because the Hunter fleet was meanwhile depleted to a degree where not enough aircraft were available for two squadrons. Moreover, there were delays with the deliveries of MiG-23s ordered in 1972 which were planned to replace the Hunters. The solution was to concentrate all the remaining Hunters within No. 6 Squadron and press the 12 'new' MiG-19s into service with No. 29 Squadron.

Pilots of No. 29 Squadron, with one of their 'new' MiG-19s, seen in 1975, shortly before the type was finally withdrawn from service and replaced by MiG-23BNs. (via Ali Tobchi)

A group of pilots of No. 29 Squadron, with the new squadron crest – including a swordfish – seen in 1974. (via Ali Tobchi)

Following several months of conversion training, the unit was declared operational in August 1974, and promptly deployed to fight Barzani's Peshmerga. Pilots and ground crews of the squadron knew that their 'new' mounts had a year of operational service left in their engines at most, but they did their best to keep them operational. Indeed, after pulling some strings, the General Headquarters (GHQ) of the Iraqi Armed Forces in Baghdad received help from Cairo in the form of stocks of spares left in local depots. Ironically, the 'new' Iraqi MiGs were thus given a new lease of life with the help of spares recovered from former Iraqi MiG-19s that were donated to Egypt in 1964. This not only bolstered the serviceability of No. 29 Squadron's jets but raised spirits to a degree where a new squadron insignia was developed, including a spearfish superimposed over the Iraqi national flag (as shown in the colour section of Volume 1).

and safe bases and had done so since the mid-1960s. By 1969, related Iranian activities reached such proportions that as 'revenge', Baghdad attempted to enforce the Saabad Treaty of 1937, and assert its control over the Shatt al-Arab waterway, in turn threatening to block access to the crucial Iranian oil-exporting ports of Abadan and Khorramshahr. The Shah quickly abrogated the treaty and deployed his armed forces to escort tankers moving up and down the waterway. Feeling too weak to fight Iran, Baghdad quietly lessened tensions. Nevertheless, both the Iranian and Iraqi armed forces were mobilised and deployed along the border once again in 1970, for the very same reason: the Shah's support for the KDP. Although no war erupted, and most troops were subsequently withdrawn, border clashes became a norm over the following years and their number and intensity grew significantly. At least 23 were reported in 1972, and another 60 a year later. Initially, most of these consisted of sniping, but gradually, artillery duels followed, and on 10 February 1973 Tehran reported a pitched battle including tanks and artillery between Badrah in Iraq, and Mehran in Iran, resulting in 41 killed and 81 wounded Iranian military personnel. Certainly enough, neither side ever cared to release precise details and tensions then lessened by an order of magnitude during the October 1973 Arab-Israeli War. However, no sooner had the dust begun to settle after that conflict, the Shah was back to the business of destabilising the 'radical regime' of Iraq through supporting Barzani's Peshmerga; indeed, he not only renewed, but greatly expanded the military support by deploying an equivalent of a brigade of regular troops of the Imperial Iranian Army, dressed in Kurdish garb, to advise and then to fight for the Kurds. As could have been expected, the GMID soon got wind of these developments, and requested that the IrAF fly reconnaissance. Using Hunter F.Mk 59s of No. 6 Squadron – two of which were re-equipped with 'reconnaissance noses' containing British-made Vinten cameras – its pilots flew about a dozen reconnaissance sorties not only over Kurdish-controlled parts in northern Iraq, but over Iran as well. Shocked by the resulting intelligence, the government in Baghdad was concerned enough to request the United Nations (UN) to deploy a peacekeeping force along its border with Iran.[1]

Before the UN could react, and in an attempt to distract Tehran's attention, Bakr and Saddam ordered their army to capture the Iranian border town of Mehran, and thus secure a bargaining chip for negotiations. On 10 February 1974, a brigade of the 3rd Armoured Division – considered the best unit of this kind in the Iraqi Army – crossed the border and approached the town. The Iraqis ran straight into an ambush. Informed about the concentration of enemy troops along the border by the intelligence-gathering activities of their top intelligence agency – the Intelligence and Security Organisation of the Country (best known by its abbreviation in Farsi: 'SAVAK') – and clandestine reconnaissance flights of McDonnell Douglas RF-4E Phantom IIs of the Imperial Iranian Air Force (IIAF), the Imperial Iranian Army was on alert. It hit back with all available means, including US-made M60 main battle tanks, BGM-71 TOW anti-tank guided missiles, and M109 self-propelled howitzers, knocking out dozens of vehicles, and killing, according to reports from Tehran, over 50 Iraqi troops. To add salt to the wound, the IIAF then deployed eight F-4E Phantom II fighter-bombers armed with brand-new AGM-65A Maverick electro-optically guided air-to-ground missiles to destroy scores of Iraqi military vehicles. Still deeply impressed by 'mighty' F-4 Phantoms from the recent war with Israel, the IrAF completely failed to rise to the challenge.

THE SHAH'S COUNTERSTRIKE

Although his armed forces proved vastly superior in the clash near Mehran, the Shah of Iran was not to let Iraqi provocation go unanswered. Knowing that the government in Baghdad had announced a new legislative for Kurdish areas, set to be enforced in April 1974 – and alongside which these areas were to receive an elected legislative assembly based in Erbil, but presided over by a person appointed by the president of Iraq, who was to have the final say in all affairs and thus secure Iraq's control – he encouraged Barzani to act. Of course, the Kurdish leader was eager to follow advice; he was not only displeased by the announced legislation, but strongly disagreed with the idea of the resulting 'Kurdish autonomous region' excluding about one-third of what he claimed for himself, foremost the oil fields in the Kirkuk area. Therefore, on 11 March 1974, a force of 25,000 Peshmerga assaulted the Iraqi Army garrison in Zakho.

Knowing it would take time to mobilise and deploy the ground forces, Baghdad ordered the IrAF into action. Within 48 hours, about 50 Hunters, MiG-15s, MiG-17s, and MiG-21s were redeployed to reinforce Su-7BMK and Su-20 units home-based at Hurrya AB, in Mosul. Just like in Syria during the previous October, their pilots initially flew reconnaissance sorties, before launching first air strikes on 14 and 15 March. Three additional squadrons of Hunters and MiG-21s, and even most of the Jet Provosts of the Flight Leaders School were to follow. During the next week they flew up to 100 combat sorties a day, foremost against Kurdish positions in the Zakho area, buying time for the army to deploy no less than eight divisions – a total of about 100,000 troops supported by about 1,200 tanks and other armoured vehicles – around Mosul, Kirkuk and Erbil, and launch its (by that point in time) largest-ever offensive against Barzani's guerrillas. The heaviest hit, apparently on Saddam's order, was the town of Qaladiza, the seat of the University of Slemani. At 0915 hours on 24 April 1974, this was subjected to air strikes by several formations of Su-7BMKs and Su-20s, using OFAB-250 cluster bombs filled with incendiaries; according to KDP reports, about 130 (later increased to 'more than 350') civilians were massacred, and over 400 wounded. Two days later, the IrAF applied similar 'aerial policing' to the town of Halabja (also 'Halabcheh'), causing scores of additional casualties. Under pressure, the KDP pleaded for help and the Shah replied positively; not only were additional shipments of firearms, mortars and ammunition rushed to the KDP, but the Imperial Iranian Army moved at least a battalion of light artillery inside Iraq, followed by a complete artillery brigade deployed along the border. All three units began providing indirect fire support for Kurdish insurgents.[2]

Four pilots of No. 11 Squadron, seen in front of one of their MiG-21MFs (serial number 1099), fresh back from deployment to Syria, in late 1973. Only six months later, they were flying combat sorties again, this time against the Kurds in northern Iraq. (Tom Cooper Collection)

When the IrAF was deployed to strike back against Barzani's Peshmerga that had attacked the Iraqi Army garrison in Zakho, it even sent Jet Provost advanced trainers of the recently established Flight Leaders School into action. This example (serial number 619) was photographed prior to delivery to Iraq in the mid-1960s. (Albert Grandolini Collection)

A row of MiG-21s at Rashid AB in around 1974–1975. Notable is the 'mix' of two MiG-21PFMs without camouflage (centre), and (from left to right), MiG-21MFs, MiG-21FLs and MiG-21PFMs with camouflage colours, some of which were applied during the deployment of Nos. 9 and 11 Squadrons in Syria during 1973. (Tom Cooper Collection)

IRAF LOSSES

The fighting in northern Iraq went on through the entire spring, summer and then the autumn of 1974, without either side achieving a major breakthrough. Initially at least, General Command in Baghdad took great care that the air force avoided striking Iranian positions. However, through autumn 1974, the IrAF began losing aircraft over northern Iraq; one after the other, single MiGs and Sukhois never returned to their bases. The air force did lose a few aircraft in fighting against the Kurds in the 1960s, but never more than one or two per campaign, and the reasons were always known. The first known loss that occurred during the fighting against Barzani's Peshmerga was a similar case; on 1 July

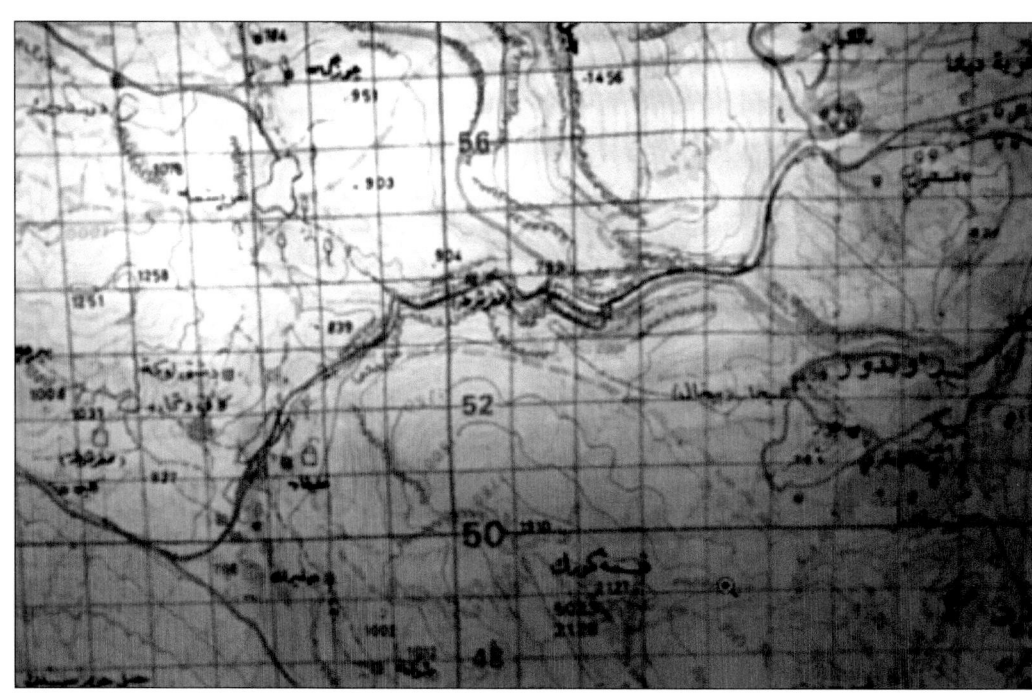

A screengrab from a video showing an Iraqi military map of the area along the road from Erbil via Salahuddin, Shaglawa, to Khalifan, where the 22nd Infantry Brigade of the Iraqi Army punched through the Gali Ali Bey Gorge, only to have its 3rd Regiment cut off after securing Mount Nawakhin. (via Tom Cooper)

1974, a brand-new Su-20 from No. 1 Squadron disappeared over the Sinjar area, together with its pilot.³ A few days later, a Hunter F.Mk 59 flown by Major Abdal Latif Abdul Karim from No. 6 Squadron was ripped apart by the premature detonation of two Spanish-made EXPAL 500kg bombs while attacking a target in the same area. However, on 18 August 1974, another jet from this unit piloted by Captain Hashem Qaddo Abbas 'Tahan' al-Obeidi – a veteran of the October 1973 War with Israel – disappeared without a trace while underway over the Kalala area.⁴

The fighting reached its peak in mid-September 1974, when the 22nd Infantry Brigade of the Iraqi Army, punched through the Peshmerga defences in the Gali Ali Bey Gorge, and secured the peak of Mount Nawakhin (also 'Chia-y Bradost'), northwest from Ruwanduz. The Kurds reacted with counterattacks into the flank of the Iraqi advance, eventually managing to cut off the 3rd Regiment of the 22nd Brigade, entrenched high on Mount Nawakhin. Determined to save the surrounded unit, the High Command in Baghdad deployed a task force including four regiments of commandos to re-open the road from Erbil with Salahuddin, Shaqlawa and Khalifan, and a regiment of engineers to construct a bypass road over Mount Korek. Both efforts received intensive close air support, with IrAF using all available fighter-bombers to keep the enemy at bay, and its transports to keep the ground units resupplied with ammunition and food. On 27 and 28 September 1974, Hunters, MiG-15s, MiG-17s, MiG-19s, Su-7s and Su-20s bombed and strafed Peshmerga along the Hamilton Road, connecting Erbil and Ruwanduz with Penjwin in Iran. Furthermore, An-12s of No. 23 Squadron attempted to support the 3rd Regiment by dropping supplies by parachute. The latter task proved particularly problematic because the local terrain and weather made it virtually impossible for transports to actually hit the peak of Mount Nawakhin and thus the positions of the surrounded unit. Therefore, the transports were soon loaded with bombs instead, and deployed to saturate insurgent positions with salvoes of FAB-250M-54s released over their rear loading ramp. While the 3rd Regiment was eventually resupplied by helicopters and held out in its positions until the spring of 1975, the IrAF suffered its next loss when a MiG-19S from No. 29 Squadron, flown by 1st Lieutenant Safa Shallal, disappeared. This time, not even the GMID had any useful explanations; the Iraqis knew that Barzani had been supported by the Soviet Union at earlier times, but all such links should have been cut off since the signing of the Iraqi-Soviet Friendship Treaty. Thus, the Peshmerga were not known to have air defence systems capable of regularly downing Iraqi fighter-bombers.⁵

DEATH VALLEY

As IrAF losses continued to mount, the situation reached a point where the IrAF began deploying Tu-16 bombers from No. 10

Once again, the IrAF went into 'all-out mode' during the fighting for Mount Nawakhin, where An-12s of No. 23 Squadron not only dropped supplies to the besieged 3rd Regiment, but also acted as makeshift bombers. (Albert Grandolini Collection)

Mi-4s of No. 2 Squadron played the crucial role in keeping the 3rd Regiment resupplied, enabling it to survive a month-long siege on Mount Nawakhin until the spring of 1975. This photograph shows one of the helicopters of this unit (serial number 678) that retained its 'civilian' livery. (via Ali Tobchi)

A still from a video showing a – then brand-new – Tu-22 bomber of IrAF, unleashing a stick of training bombs. (Tom Cooper Collection)

Squadron; on 27 and 28 September 1974, these bombed Barzani's headquarters in the mountains above Haji Omran, a town in the Choman District situated at an altitude of 3,000 metres (9,843 feet). Bombers could do so from high altitude, where they were considered safe from any kind of air defences, and with the help of massive FAB-1500 and FAB-3000 bombs (1,500kg/3306lbs and 3,000kg/6613lbs, respectively). The latter tended to cause damage even if failing to score direct hits. However, on 15 December 1974, a Tu-16 bomber piloted by Captain Khalil Abud Hassan al-Musawi disappeared while underway over the Khoshkan area, close to the Iranian border, together with all six crew members. Combined with reversals experienced by the army because of the Iranian artillery, this prompted the General Command in Baghdad to order the IrAF to find and hit back against the Iranians. On 16 December 1974, more than 40 Hunters, MiG-15s, MiG-17s, MiG-19s, Su-7s and Su-20s flew the first air strikes on Iranian artillery positions. It was only during this operation that the Iraqis finally realised what were they facing: a MIM-23B I-HAWK SAM site of the Iranian Air Force was detected when it opened fire and shot down one of the Su-7BMKs. A day later, another IrAF fighter-bomber was felled by an I-HAWK only a few kilometres short of the Iranian border, prompting the air force to stop all further operations within areas known to be protected by SAMs. From that time onwards, the powerful US-made SAM-system became known as 'Death Valley' in the Iraqi Air Force.[6]

Instead of fighter-bombers, the General Command now ordered the IrAF to bring in its Tu-22s. These were armed not only with FAB-1500s and FAB-3000s, but with more massive FAB-5000 and FAB-9000s (5,000kg/11,023lbs and 9,000kg/19,842lbs, respectively). By releasing them while flying the so-called 'supersonic-toss manoeuvre' – a manoeuvre even their instructors in the USSR considered much too dangerous to fly with the Tu-22 – the crews were able to avoid flying into the centre of the engagement envelope of the Iranian SAMs. Furthermore, the Iraqis realised that aiming at snow-covered mountain peaks was more effective than aiming at their actual targets: the massive blast from their bombs usually caused an avalanche of snow and debris that destroyed anything on the way down. Throughout December 1974, the Tu-22 crews of Nos. 18 and 36 Squadrons flew so many air strikes in this fashion that they completely exhausted the available stocks of FAB-1500s, FAB-3000s, FAB-5000s and FAB-9000s. Not one Tu-22 was hit in return. Of course, the Ministry of Defence in Baghdad rushed to order replacement weapons; quite to the surprise of everybody involved, Moscow turned down all such requests.[7]

ALGIERS ACCORD

With the fighting reaching a stalemate, and unknown to Barzani, Baghdad and Tehran entered secret negotiations, mediated by the Algerian government. The outcome was the dramatic announcement of the Algiers Accord, signed by Saddam Hussein and Shah Reza Pahlavi on 6 March 1975, which served as the basis for bilateral treaties concluded on 13 June and 26 December of the same year. Mostly due to Saddam Hussein's initiative, Iraq accepted Iran's claim that the thalweg (deepest channel) should form the common boundary of the two states in the Shatt al-Arab. In exchange, Tehran promptly ceased all support for Barzani's forces. Literally overnight, all Iranian units were withdrawn back across the border. The Kurdish resistance collapsed within the next two weeks and thousands of Peshmerga surrendered, while the leadership of the KDP fled into Iran, together with about 150,000 civilians. Moreover, festering divisions within the leadership of Barzani's movement led to a major split between Mulla Mustafa and his sons Idris and Masoud, and Jalal Talabani. The latter broke with the KDP and established the Popular Union of Kurdistan (PUK), drawing with him all those in disagreement with Barzani's tribal leadership, which they found hard to reconcile with their nationalist and leftist principles.[8]

It remains unknown exactly how many combat and combat-support sorties the IrAF – which was expanded from 9,800 to 10,500 officers and other ranks during 1974 – flew during the Second Iraqi-Kurdish War. However, available reports indicate an overall effort even larger than that during the October 1973 Arab-Israeli War. Losses were at least significant: Barzani's Peshmerga claimed a total of 18 Iraqi aircraft as shot down, including two Tu-16 bombers, two Hunters, at least one Su-20 – and even several 'MiG-23s', which actually did not participate in operations. Obviously, most of these were actually hit by MIM-23B I-HAWK SAMs operated by IIAF-personnel, disguised as 'Kurdish insurgents'.[9]

With the end of this conflict, Baghdad was free to press ahead with its plan for the Kurdish region. A Kurd, Taha Muhi ad-Din Ma'ruf, was appointed Vice-President of Iraq, while the armed forces launched a major operation of relocating entire Kurdish communities – between 400,000 and 500,000 people – in a southern and eastern direction, away from the borders with Iran and Turkey,

From left to right: Shah Mohammad Reza Pahlavi, President of Algeria Houari Boumediene, and Vice-President of Iraq, Saddam Hussein, in Algiers, on 6 March 1975. (Albert Grandolini Collection)

often even into predominantly Shi'a populated parts of southern Iraq. In turn, the government encouraged Arab families to move northward, and initiated several major construction projects, primarily related to roads and electricity, making sure the north would become better accessible to their own armed forces.

4
DULAYMI'S TROUBLED TENURE

For Bakr and Saddam, there were two primary lessons from the two wars Iraq fought in 1973–1975, and the Algiers Accord (which Saddam subsequently described as a 'shame'): Iraq had to greatly expand its armed forces and it had to find alternative sources of arms. The first related discussions had been held already in January 1974, when the performance of the Iraqi armed forces during the October 1973 War was discussed during the Eight Ba'ath Party Congress in Baghdad. By this time, Baghdad was yielding immense profits from oil exports and the so-called oil crisis of 1973, caused by the members of the Organisation of Arab Petroleum Exporting Countries – led by Saudi Arabia – proclaimed an oil embargo upon all countries supporting Israel, including Canada, Japan, the Netherlands, Portugal, Rhodesia, South Africa and the USA. As the price of oil jumped by nearly 300 percent, from a mere US$3 per barrel, to US$12 per barrel, globally, so did Iraq's income; it jumped to more than US$2 billion in 1974, and then continued to grow by 40 percent per annum for the rest of the decade. With Iraq's economy experiencing a period of high growth and soaring prosperity, in 1976 the RCC launched a major expansion of the armed forces. Just as the manpower of the army was doubled over the following two years, so too was the IrAF to experience a quantitative and qualitative growth of unprecedented dimensions. Indeed, while traditionally equipped and trained to preserve internal security and run counter-insurgency (COIN) campaigns, by 1980 the Iraqi armed forces became the most powerful military of the Arab world, converting Iraq into a regional power.[1]

PROJECT 202[2]
One of the most obvious results of Iraqi experiences from both the wars with Israel in 1967 and 1973 was the decision to not only harden available air bases, but also order the construction of new, extensive and heavily fortified facilities for the air force. A large number of blast pens had already been constructed by British companies on existing airfields by 1973. The MIC then learned about the construction of a major underground facility in Yugoslavia, and thus sought connections to Belgrade.[3] The result was negotiations that led to a contract for Project 202 – the construction of five new air bases – in 1975. Designed by a team of Yugoslav engineers with fresh experience from constructing extensive underground facilities at Bihac Air Base, Project 202 – sometimes known as 'Super Base' – included the reconstruction of three existing airfields

Older IrAF air bases – like Firnas, in Mosul, visible in this photograph from November 2003 – had received groups of double hardened aircraft shelters constructed by British companies in the early 1970s. (Photo by Martin Rosenkranz)

Aerial view of Bakr AB (better-known in the West as 'Balad'), constructed to the 'improved trapezoid' design: note adjacent to the end of each runway are up to seven hardened aircraft shelters, each with its own taxiway. (Photo by Martin Rosenkranz)

Aerial view of Abu Ubayda Ibn al-Jarrah AB, outside al-Qut, as seen in November 2003. Clearly visible are two runways, and a total of five 'groups' of four hardened aircraft shelters near the western and south-western ends of the main runway. (Photo by Martin Rozenkranz)

– H-1 (202A), H-2 (202B), and H-3 (202C) – and the construction of two entirely new air bases: Bakr (outside Balad), and Saddam (outside Qayarah, north of Tikrit). Each of these facilities was to cover an area of around 40 square kilometres (21.5 square miles), have two concrete runways each about 3,800 metres long (12,420 feet) and 60 metres (200 feet) wide, and to have a large number of intersections and taxiways that could be used as runways in case of emergencies. Other installations on each of the Project 202 bases were divided into two categories: surface and underground. Where the underground work was limited by the local terrain, because of the relatively high level of water and oil under the local soil, instead of burying major facilities under the ground, the Yugoslavs had to make extensive use of reinforced concrete, only partially buried into the surrounding terrain. The primary constructions were over 200 hardened aircraft shelters on these five bases alone: each was wider than similar objects constructed around the same time in Eastern or Western Europe and could house even the largest tactical combat aircraft then in service. Every hardened aircraft shelter consisted of a one-metre-thick concrete wall, reinforced both inside and out by 30mm thick steel plates. The front entrance was covered by sliding doors made of 50cm-thick steel armoured plate. Such shelters were usually constructed in groups of four to six, with each group sharing the same water and power supply; however, every shelter received its own backup gasoline-powered electrical generator. Ammunition depots were constructed in a similar fashion, that is, partially buried, with their sides and roofs of steel plates, with concrete in between. Offices and squadron ready rooms were made of reinforced concrete and constructed on the surface, while maintenance hangars were metal structures covered by corrugated planking – sufficient to protect aircraft and ground crews from the elements during peacetime operations, but not from damage in the event of war. That said, all the major parts of every air base were connected with a network of underground corridors, all buried as deep as 50 metres (165 feet) under the ground; just like their command centres and shelters for personnel, they were hardened to withstand anything but a direct nuclear blast.

The biggest of the five Project 202 bases became 202C; al-Wallid AB. This not only received two spaced runways and about two dozen hardened aircraft shelters (including several which were for pairs of large tactical combat aircraft), but also three dispersal facilities were constructed nearby. The Wallid Northwest and Wallid Southwest facilities each had the capacity to house an entire squadron of combat aircraft inside four hardened aircraft shelters constructed near the end of the runway, each of which had two entrances protected by tall concrete walls reinforced by earth. All three air bases, plus the nearby al-Wallid Highway dispersal site, also had underground storage facilities for ammunition and fuel. As such, the new complex could support operations of up to six squadrons of aircraft in the event of another war with Israel.

Further east was Project 202B, the former H-2 airfield, which received a new runway with about a dozen hardened aircraft shelters. However, the centrepiece of this project became the H-1 airfield (202A), which was reconstructed in a fashion similar to 202B, but which received a separate taxiway connecting the end of the runway to every single one of about a dozen hardened aircraft shelters constructed nearby. The resulting design became known as 'Trapezoid' in the West and was to strongly influence the construction of al-Bakr AB and about a dozen other air bases subsequently constructed in Iraq. Because of the high satisfaction with the Project 202, this encouraged the MIC into contracting the Yugoslavs to build three additional major facilities for the IrAF: Ubayda Ibn al-Jarrah, outside al-Qut; al-Iskandariyah, south of Baghdad; and Ali Ibn Abu Talib, south of an-Nasiriyah. The first two were surrounded by intensively cultivated areas, and thus lacked space for more comprehensive construction of hardened facilities. Correspondingly, each received three groups of four hardened aircraft shelters at either end of the runway. However, there was enough space at the latter site and thus – just like al-Bakr before – Ali Ibn Abu Talib AB became a further development of Project 202A, though it was expanded to offer enough space for a full wing of three squadrons of tactical combat aircraft each. All four of these major new air bases – al-Wallid, al-Bakr, Ubayda Ibn al-Jarrah, Iskandariyah and Ali Ibn Abu Talib – became operational by 1977–1978, and were to serve as major hubs in future operations of the IrAF.

DISAPPOINTING MIG-23[4]

The first of the newly constructed or reconstructed air bases reached their initial operational capability just in time to receive some of the new aircraft ordered by Baghdad in reaction to experiences from the October 1973 War. Once again, the Ministry of Defence and the GHQ in Baghdad, and the Chief of Staff IrAF, Major General Dulaymi, did little more than repeat the earlier exercises – which was to replace losses through the acquisition of yet additional aircraft from the USSR. In October 1973, the Soviets delivered the first 10 Su-20s (as mentioned previously); completed by eight additional examples that followed in early 1974, these were introduced to service with No. 1 Squadron as the unit returned from Syria. During the same period, the losses of the MiG-21s of No. 9 Squadron were replaced by an additional batch of 18 MiG-21MFs; of particular interest was that the latter arrived together with a batch of brand-new R-13M-1 air-to-air missiles, with a much improved performance and a wider engagement envelope in comparison to the older R-3S, which had proved near-useless during the October 1973 War.[5] However, subsequently the IrAF intended to re-orient and replace this type with an entirely new Soviet-made interceptor: the MiG-23. This issue was of growing urgency because not only was the IIAF in the process of acquiring over 100 additional F-4E Phantoms, but especially because in November 1973 Tehran placed an order for 80 brand-new, powerful Grumman F-14A Tomcat interceptors for deliveries starting in early 1976.

As indicated earlier, the Iraqi plan to acquire MiG-23s soon began falling apart due to delays in delivery; even once the aircraft did begin to arrive, the entire project nearly collapsed because of their poor manufacturing quality and even poorer 'customer service' by the Soviets. In late 1973, a group of selected Iraqi MiG-21 pilots travelled to the USSR for conversion training to the new type. On arrival they were startled to find out that, although none had less than 1,500 hours of experience on fast jets alone, the Soviets treated them as novice cadets. Out of the next six months of training, four-and-a-half were spent on the ground: the actual flying time was minimal, and limited to take-offs, flying circles around the base, and landings. Pilots that hoped to be trained in flying MiG-23s to the edge of their envelope thus came back home with a sour feeling. What followed once the first aircraft arrived in Iraq, in summer 1974, was a shock. The first variant delivered – MiG-23MS – was supposed to be an interceptor praised by the Soviets as 'superior to the F-14'. Actually, it was equipped with the weapons system from the MiG-21bis (including an obsolete radar, four R-3S missiles and a GSh-23L twin-barrel 23mm automatic gun) and it proved to be a technological catastrophe, not only poorly manufactured and thus prone to malfunctions, but possessing very narrow aerodynamic limitations. Moreover, the Soviets delivered it with only partial technical documentation, and a poorly translated flight manual. Former MiG-21 pilots, accustomed to flying an almost vice-free jet under all conditions, were facing unheard-of limitations when trying to operate the MiG-23MS: the result was a series of accidents – several of these fatal – that rapidly depleted the fleet of brand-new aircraft. The crisis reached such proportions that the planned conversion of No. 7 Squadron to this type had to be abandoned. The unlucky unit had to be disbanded once its surviving MiG-17Fs and MiG-17PFs were withdrawn due to the exhaustion of their airframes in 1975.

Meanwhile, in 1975 and 1976, the IrAF received the first batch of 18 MiG-23BNs. This variant was custom-tailored for ground attack

A brand-new Su-20 (serial number 1303) of No. 1 Squadron, around 1974–1975. Notable is the installation of four hardpoints under the centre fuselage. (Ahmad Sadik Collection)

A still from a documentary film offering an elevated view of one of the MiG-23MS delivered to Iraq in summer 1974. (Tom Cooper Collection)

A profile of an Iraqi MiG-23MS following its upgrade through re-wiring for the installation of R-13 air-to-air missiles on underwing stations. Sadly, the serial and the unit insignia of No. 39 Squadron have both been blotted out by the Iraqi military censor. Notable is the white radome – as on many of the early Iraqi and Syrian MiG-23MS. (Tom Cooper Collection)

and thus lacked the radar. Instead, it had a 'duck-nose', designed to offer the pilot a good view to the front and below. As planned three years earlier, the new aircraft entered service with No. 29 Squadron, enabling it to, finally, withdraw from service its old MiG-19S, and move to the newly constructed Ali Ibn Abu Talib AB in the desert south of an-Nasiriyah. The second batch of 18 MiG-23BNs followed a year later and entered service with the newly established No. 49 Squadron, based at another newly constructed air base: Abu Ubayda Ibn al-Jarrah AB, outside al-Qut ('al-Kut').[6]

The MiG-23BN was not designed for air combat, and was thus rarely flown to the edge of its envelope; nevertheless, it suffered from the same poor production quality and lack of associated technical documentation like the MiG-23MS. Worse yet, while the Iraqi order – and the related contract – explicitly demanded delivery of aircraft equipped with SPS-141 ECM-systems and internal Delta guidance installations for semi-automatic Kh-23M radio-command guided air-to-ground missiles (ASCC/NATO-codename 'AS-7 Kerry'), all the first 36 MiG-23BNs arrived lacking such systems – despite Baghdad paying for these. The Soviets also refused to train Iraqis in the use of their pre-programmable navigation systems.[7]

Eventually, the status of the Iraqi MiG-23 fleet improved though only very slowly, with pilots 'learning by doing'. No fewer than 12 MiG-23MS, MiG-23BNs, and MiG-23UBs were lost in flying accidents by 1978: by then, the IrAF issued a completely new flight manual, based on flight-testing in Iraq. Only then was the MiG-23MS declared operational with the newly established No. 39 Squadron, based at Wallid AB in western Iraq.

SUKHOI MAFIA[8]

In the meantime, the government of the USSR signalled its preparedness to continue selling Su-20s to Iraq. However, now even Dulaymi and other IrAF officers that preferred Soviet-made fighter designs to Western – colloquially known as the 'Sukhoi Mafia' within the air force – could not ignore the negative experiences with MiGs and Sukhois during the recent wars with Israel and the Kurds. Unsurprisingly, they began making ever-more stringent demands: if the IrAF was to continue purchasing Sukhois, then Moscow was to deliver their best-equipped variants. The result was protracted negotiations that postponed the next order to 1976; this secured the delivery of 36 Sukhoi Su-22 fighter-bombers. Outwardly similar to Su-20s, the new jets were equipped with a comprehensive avionics suite, including a Doppler speed and drift sensor installed under the chin, compatibility with Kh-23 missiles and large KKR-1 reconnaissance pods that could be installed under the centre fuselage. They entered service with two newly established units, Nos. 44 and 109 Squadrons, both staffed by a combination of experienced Su-7 pilots and novices from the Air Force Academy. Much to the disappointment of the Sukhoi Mafia, the Iraqis found the Su-22s unsatisfactory; exactly like the MiG-23BNs, these arrived without the internal equipment necessary for the deployment of Kh-23M missiles. Instead, the Soviets delivered only the original variant of this weapon (designated Kh-23E for export purposes), which quickly proved far too cumbersome for deployment from a single-seat fighter-bomber under combat conditions. This was the virtual drop that over-spilled the barrel and Dulaymi reacted with an ultimatum: either Moscow would take his requests seriously, or he would opt for Western designs instead. Once again, the Soviets took a long time – and then an actual Iraqi order for a Western-made fighter-bomber, plus a new Chief of Staff IrAF – to make up their minds.

POINT DEFENCES FOR STRATEGIC INSTALLATIONS

Although deciding not to follow the Egyptian and Syrian emphasis on complex networks of surface-to-air missiles for the first line of defence, the Iraqis understood that manned interceptors should not be the sole means of air defence, and that air bases and major civilian and military facilities – generally considered 'strategic installations' by the Iraqis – required a 'second line of defence'. Correspondingly, through the first half of the 1970s they continued purchasing Soviet-

A V-750VMV (also '20D') missile of an Iraqi S-75M Volkhov (SA-2) SAM-system in position. In the IrAF, such weapons were operated by dedicated missile brigades, each of which included a mix of up to six SA-2 and SA-3 battalions and was responsible for point defence of most important air bases or major civilian facilities. (Tom Cooper Collection)

made SAM systems, even if in very limited amounts. For example, at the time Egypt and Syria were acquiring dozens of such systems every year, Iraq bought only four S-75 Dvina systems (that is, enough equipment for four battalions or four 'SAM sites') which were purchased and delivered in 1974, three in 1975 and four each of the improved S-75M Volkhov in 1976 and 1977 – together with no fewer than 570 associated D-20, and 10 V-750VMV missiles (ASCC/NATO-codename for all these systems was 'SA-2 Guideline'). During the same period, the IrAF purchased 24 S-125 Pechora (ASCC/NATO-codename 'SA-3 Goa') SAM systems, together with 850 associated V-601 missiles. On top of this, Baghdad purchased a total of eight P-14 and P-14F long-range early warning radars (ASCC/NATO-codename 'Tall King'), with the maximum detection range of 400 kilometres (250 miles) and up to an altitude of 30,000 metres (98,000 feet). As usual, the Iraqis heavily emphasised the training of their crews at home, and therefore also acquired 21 Akkord 75/125A simulators. When combined with the existing network of radars used for ground control of aerial operations, all linked up with a single Azurk-1ME automatic tactical management system (ATMS), acquired in 1974, and 30 Bastion-3E(F) command posts (useful for linking the work of ground-based air defences with that of manned interceptors), these systems formed the backbone upon which the IrAF eventually established the first integrated air defence system (IADS), covering the whole of Iraqi airspace, in 1976.[9]

MIC AND THE TECHNICAL DEPARTMENT[10]

Meanwhile, two other major developments related to the IrAF took place. In 1974, Saddam established the Military Industrialisation Commission (MIC, also known as the Directorate of Military Industries) and took care for this to become a major contributor to the Arab Organization for Industrialization (AOI) – an Egyptian-based Arab military organization co-sponsored by the Kingdom of Saudi Arabia, the UAE and the Emirate of Qatar, with the intention of supervising and coordinating the collective development of the pan-Arab arms industry.[11]

How important this decision was for the Iraqi strongman demonstrates the fact that he appointed his two closest (even if very young) confidantes in charge of the MIC: Hussein Kamel Hassan al-Majid and Amer Mohammed Rasheed al-Obeidi. Their task was to develop the national defence industry sector to minimise the dependence upon, and vulnerability to, external factors, but also as a leading edge for military training and the development of skills for the civilian sector, and to lower foreign exchange expenditure and dependence. Over time, related activities by Majid and Rasheed resulted in the emergence of a huge network of research centres, state enterprises, front companies and dual-use factories capable of producing both civil and military goods.[12]

The build-up of the MIC went hand-in-hand with another very important issue: prior to the October 1973 War with Israel and the Second Iraqi-Kurdish War of 1974, the Iraqis had never encountered, nor ever made use of electronic warfare (EW). Experiences from fighting the Israelis and Iranians taught them about the importance of what initially appeared to them as a 'mysterious and highly secretive method of warfare'. Consequently, through 1974 and 1975, the armed forces set up several conferences to analyse related lessons. One was that EW must become a part of their doctrine, and that there was a need to obtain suitable equipment. In 1974, each of the three branches of the armed forces had established their own Technical Departments, each responsible for the acquisition and operations of EW-equipment. From that point in time, the acquisition of high-technology became the number one priority – both for the Bakr-Saddam government, and for the Iraqi armed forces.

The IrAF started literally from scratch: initially, it had no equipment for electronic warfare at all. It quickly had to learn that such hardware was a scarce commodity even within the few advanced industrial nations of the time. Thus, the beginning was slow and hard, and it took time for

A brand-new MiG-21R of the IrAF (probably serial number 1828), seen on finals to landing at Rashid AB, in around 1977–1978. Barely visible under the centreline is the Type P ELINT reconnaissance pod. (Tom Cooper Collection)

the development of the Technical Department, IrAF, to pick up some speed. The first EW-related equipment acquired by the Iraqi Air Force was four MiG-21Rs delivered by the USSR in 1976 and integrated into the Rashid-based No. 70 Squadron. These four jets belonged to a sub-variant of the older MiG-21PFM, which was adapted to carry one of two types of reconnaissance pods. The Type-D pod carried conventional reconnaissance cameras and the equipment necessary to operate them. Based on Egyptian experience, the Iraqis quickly replaced Soviet-made cameras with British-made Vinten cameras (later on, the IrAF acquired British-made pods designed for deployment from the MiG-21R). The Type P pod was used for electronic reconnaissance (ELINT) and contained receivers for electronic emissions and recording equipment. The latter was of now-obsolete technology: essentially, it used a moving strip of photographic film, on which it could record the location of detected enemy radars (with an accuracy of 'several kilometres') and determine their working frequencies.

The four MiG-21Rs arrived together with a group of Soviet advisors; these helped work up No. 70 Squadron's plots and ground crews, and then lectured them in reconnaissance operations. Under the Iraqi doctrine of national defence in play since 1948, there were two primary enemies: Israel and Iran. With geography preventing them from running reconnaissance operations against Israel, the Iraqis focused on Iran. Therefore, for the next two years, the IrAF MiG-21Rs flew a series of reconnaissance missions along the Iranian border, and their results were then analysed by Iraqi specialists with Soviet help. The Soviets left after two years, but No. 70 Squadron continued flying ELINT-operations along the border with Iran even once it was redeployed to Wallid AB and over time, its crews managed to collect a relatively precise, even if still a rather 'shallow' and 'smudged' picture of major Iranian radar stations.

5
SHA'BAN'S FIRST TOUR

Midway through all the negotiations with Moscow, deliveries of new Soviet-made aircraft, and construction of new bases, in June 1976, the IrAF experienced its first routine change of command in two decades. Lieutenant General Dulaymi retired and was replaced by Major General Hamid Sha'ban at-Tikriti. Aside from being 'another' officer from the Tikriti clan, Sha'ban was – and remains – one of the most controversial figures in the history of the Iraqi Air Force, to the degree where nowadays he is almost completely ignored by his former comrades and Iraqi historians alike, regardless of having played such an important role in the history of his service, and of Iraq as a nation. Some IrAF veterans outright despise Sha'ban, calling him a 'Sunni extremist' and an officer responsible for 'ordering pilots into certain death'. Others, including not a few Western observers, have declared him a 'political' officer, essentially Saddam's puppet, lacking the ability to make decisions on his own, and somebody who earned his new position only thanks to affiliation with Saddam Hussein.

COMMANDER AND SURVIVOR: HAMID SHA'BAN AT-TIKRITI

Born in Tikrit, in 1931, into the al-Bu Nassir Tribe, Sha'ban joined the Engineering College in 1949, but then decided to switch to the Air Force College instead, once this was re-opened, a year later. He graduated in 1953, and was appointed to No. 5 Squadron, then equipped with de Havilland Vampire jet fighters. Having distinguished himself, he was quickly advanced to the position

Iraq was always a strong supporter of the Palestinians, and this was clearly demonstrated in much of the IrAF insignia, many squadron crests of which contained the Palestinian flag. This photograph shows Hamid Sha'ban with the leader of the PLO, Yassir Arafat, during the latter's visit to No. 9 Squadron, in 1969. (via Ali Tobchi)

of a commander of this unit before, in 1958, being reassigned to No. 6 Squadron, equipped with Hunters. Nothing is known about Sha'ban's affiliations during the turbulent 1958–1963 period, except that he underwent a staff course, and – in 1965 – was assigned to the United Arab Command in Cairo. Two years later, Sha'ban returned to Iraq and was appointed to the command of Habbaniya Air Base (and thus, a wing including two squadrons of Hunters, and one of Tu-16 bombers). In 1966, he distinguished himself while thwarting the coup attempt by Brigadier General Arif Abd ar-Razzaq and was rewarded with the rank of Lieutenant Colonel. During the June 1967 Arab-Israeli War, Sha'ban almost single-handedly ran the IrAF operations against Israel, including the integration of Jordanian pilots, and their joint operations with his two Hunter squadrons. These culminated in the air battle over the H-3 airfield on 7 June 1967 when Iraqi, Jordanian, and one Pakistani pilot claimed a total of five Israeli fighter-bombers shot down, in return for just one loss (depending on the source, the Israelis confirmed three or four losses; one way or the other, they never attacked H-3 again). Sha'ban is often named as one of the IrAF officers involved in the Ba'ath takeover in 1968, but precise details remain unknown. Certainly enough, immediately after that he was advanced in rank to Colonel and appointed to the General Command in Baghdad, thereby starting his ultimate climb to the top ranks of the air force.

Colonel Hamid Sha'ban (left) with Colonel Mohammed Jasim Hannish al-Jaboury (centre), during the visit of President Bakr (right side) to Habbaniya AB in early 1970. (via Ali Tobchi)

In June 1970, when Brigadier General Hussein Hiyawi Hamash at-Tikriti took over as the Chief of Staff, IrAF, he appointed Sha'ban his Deputy Operations. Reaching the rank of Brigadier General, Sha'ban retained this position when Major General Nimma Abdullah ad-Dulaymi took over as the Air Chief in June 1973, and thus played a crucial role in organising and directing the IrAF's involvement in both the October 1973 War with Israel and then the Second Iraqi-Kurdish War of 1974. As such, Sha'ban was best qualified for the top rank when Dulaymi retired in June 1976.

With hindsight, there is no way around the conclusion that Sha'ban was a survivor; he never lost his command during any of the numerous coups and coup attempts of the 1950s and the 1960s. Whether involved or not, he survived the 14 Teshreen Revolution of 1958; he survived the turbulent period of Brigadier General Abd al-Kraim Qasim's rule; he retained his position during Brigadier General Arif Abd ar-Razzaq's period but then opposed at least the second of two coup attempts plotted by that officer; he survived his tour of duty as the CO Taqaddum AB during and after the June 1967 Arab-Israeli War; he was not dismissed once Hamash was replaced, and he excelled in organising the mass redeployments of the IrAF both in 1973 and 1974. It is unlikely that all of this was an accident. On the contrary, it is indicative of Sha'ban as an excellent organiser and strategist. In turn, this made it unlikely that he was 'simply' one of the best qualified officers left in the air force by 1976; even if, as the following developments were to show, when Sha'ban took over, the IrAF received the best imaginable commander at that moment in time. While there is little doubt that he did have excellent relations with Saddam, precisely this made him perfectly positioned to not only suppress the influence of the Ba'ath Party apparatchiks (who had no clue about modern airpower whatsoever), but also the Sukhoi Mafia (which led exactly nowhere), and to initiate the – long overdue – fundamental reform of the IrAF. There is no doubt that Sha'ban was the Air Chief who launched the process of converting the air force from secondary service thinking and operating defensively and tactically, into a tool capable of winning entire wars in Iraq's favour.[1]

NEW IDEAS AND SOLUTIONS

Sha'ban's appointment as the Air Chief came around the time the Iraqi economy was booming. On 27 June 1976, the RCC announced a new five-year development plan, spanning the period up to 1980 and worth US$49 billion. Supported by rapidly expanding oil production and exports, this was 13 times larger than the plan from 1970 and anticipated an annual growth of the gross national product of 16.8 percent. Iraq was thus flush with money, and there were enough means and opportunities to realise numerous ideas.

Sha'ban's tour with the United Arab Command in 1965 and 1966 had exposed him to the influence of Major General Abd el-Moneim Riyadh, at the time considered the most professional and best-educated Egyptian military officer. Fresh from advanced courses in air defence in the USSR, Riyadh became a strong proponent of what in the West became known as the integrated air defence system (IADS): a system that would fully integrate the work of all available radar stations and observation posts, units equipped with manned aircraft, surface-to-air missiles and anti-aircraft artillery, into a unified system covering the airspace of either Egypt, or all Arab countries and enabling a centralised command and control of all operations. While Riyadh's idea was only slowly adopted and then realised in Egypt in the late 1960s, before becoming a 'norm' for that country and Syria, Sha'ban – thanks to his good relations with Saddam – was able to launch the work on establishing an IADS for Iraq as soon as he took over the IrAF in 1976. Indeed, he soon proved capable of following and applying additional of ideas of Riyadh's, including one for the construction of numerous additional air bases, so as to better disperse the force, making it less susceptible to a sudden attack. Finally, he prompted a reorganisation of the command system of the air force, initiated the development of new training methods and their closer integration with the Operations Department, and then played an important role in finding alternative sources for equipment and their acquisition. The overall result was the initiation of multiple acquisitions and reforms.[2]

STRATEGIC ALLIANCE WITH FRANCE

Not only for Sha'ban, but also for Saddam, the most important lesson of the Second Iraqi-Kurdish War of 1974–1975 was the one about dependence on a single source of equipment – the USSR. Both had argued that the Soviets had their own interests, which included their relations to the Kurds, and their emphasis on maintaining the 'status quo' between the Arabs and Israel in the Middle East. Therefore, they concluded that Iraq needed an alternative source of advanced technology and arms.[3]

What came to Sha'ban's aid now was the fact that two years before he took over as the new Air Chief, the Iraqi and French governments – Saddam Hussein and the French President Georges Pompidou – had entered a sort of strategic alliance, based both on the 'special' status that France meanwhile enjoyed in the Arab world following its arms embargos against Israel during 1967–1968 (clandestinely violated by Paris all the time), and the French decision in 1972 to start importing Iraqi oil despite opposition from London and Washington. At the time, the defence sector in France was fresh out of a period of dramatic reforms, which reoriented it towards the development of advanced technologies, but also towards closer cooperation with export customers. To better coordinate such activities with requirements of its own armed forces, in 1965 the government in Paris established the Ministerial Delegation for Armament (*Délégation Ministérielle pour l'Armement*, DMA). The DMA soon proved influential during the work on the next generation of French-made combat aircraft when, together with leading local

Vice-President of the RCC, Saddam Hussein with President of France, Georges Pompidou, during Saddam's first visit in France on 14 June 1972. France's decision to become a major customer for Iraqi oil exports, despite fierce pressure from Washington and London, created a basis for close cooperation with Baghdad that was to last for nearly 20 years. (INA)

aerospace corporation, Avions Marcel Dassault–Bréguet Aviation (AMD-BA), it prompted the French air force to abandon the development of a large and complex interceptor, and instead led it to opt for a smaller and cheaper airframe with multi-role capabilities. The result was the AMD-BA Mirage F.1C: originally a medium-altitude interceptor armed with up to two Matra R.530 missiles, and two DEFA 30mm automatic guns. As soon as the new jet was ready for series production, the DMA's Direction of International Affairs (*Direction des Affaires Internationales*, DIA) – its office for export business – launched an aggressive advertising campaign abroad, attempting to attract orders. In April 1974, only months after border clashes between Iraq and Kuwait, the DIA offered the Mirage F.1 to both the governments in Baghdad and Kuwait City. With success: the result was perhaps the most important arms deal involving Iraq of the 1970s and 1980s – the acquisition of fighter-bombers made in France.

PROJECT BAZ

Arguably, having the background of being British-trained pilots that used to fly Hunters, both Sha'ban – and Dulaymi before him – would have preferred the acquisition of the McDonnell Douglas/British Aircraft Corporation F-4K Phantom II which was the variant developed for the requirements of the Royal Air Force (RAF), and the Fleet Air Arm of the Royal Navy. This was a large and powerful fighter, capable of carrying 3,000kg of bombs at a speed of more than 1,000km/h (540 knots), while still carrying air-to-air missiles for self-defence. It had been deployed intensively by the Israeli air force to strike even the most distant targets in Egypt and Syria of the 1968–1973 period, and it left deep impressions in the Arab world. However, the government in London flatly refused any kind of related talks. This was the opportunity Paris was waiting for. On 2 December 1974, French Prime Minister Jacques Chirac paid a visit to Baghdad to court Saddam, his influential cousin Adnan Khayrallah at-Talfah, and Air Chief Dulaymi. Indeed, he went as far as to invite the Iraqis to test-fly Mirage F.1s and another new fighter-bomber, the Franco-British SEPECAT Jaguar. In early 1975, Dulaymi thus dispatched Lieutenant Colonel Adel Suleiman to do so at Istres AB in France, resulting in a very positive report – at least about the first of the two types. The fact that the Jaguar was developed and manufactured in cooperation with Great Britain, and London continued complaining about the good relations between Baghdad and Paris, in fact disqualified it both in French and Iraqi eyes. After Saddam made a major visit to France in September 1975 – accompanied by the Chief of Staff Iraqi Armed Forces, Lieutenant General Abd al-Jabber Shanshal, – to see a demonstration of the new jet with his own eyes, two months later, on 21 November 1975, Saddam made the decision to buy Mirage F.1s. While the related contract was to be negotiated by his aides, and as an introduction to the cooperation with Paris, he placed a huge order for 40 Aérospatiale SE.316C Alouette III armed with 264 AS.12 anti-tank guided missiles (ATGMs), 40 SA.342 Gazelle helicopters armed with 360 advanced HOT ATGMs, nine Aérospatiale SA.321H and 5 SA.321GV Super Frelon helicopters, of which the latter were armed with 60 of the latest Aérospatiale AM.39 Exocet anti-ship missiles.[4]

The reason for the prolonged negotiations for the Mirages was that the Iraqi Vice-President – who meanwhile had himself promoted to the rank of Lieutenant General on the basis of an honorary degree from the Iraqi Army Staff College – was not only keen to obtain aircraft, but for the French to develop entirely new, custom-tailored armament for Iraq, and then to effect a full technology transfer: this demand became even more important once Sha'ban was appointed the Air Chief in 1976. Because of this, and despite a mutual agreement between the French and Iraqi governments, related talks went on for months and then years, and it was only on 3 July 1977 that Baghdad signed the contract for Project Baz ('Falcon' in Arabic) with a consortium of French companies presided over by Thales, worth US$1.8 billion.[5]

ELECTRONIC WARFARE SYSTEMS

The centrepiece of Project Baz was to become the delivery of four Mirage F.1BQ two-seat conversion trainers, 16 Mirage F.1EQ interceptors, and, later on, 16 slightly improved Mirage F.1EQ-2 fighter-bombers, equipped with the Thomson-CSF Cyrano IVQ Ramadan radar, SNECMA Atar 9K50 engines and underwing pylons, two flight simulators, 267 Matra Super 530F and 534 Matra R.550 Magic air-to-air missiles, and a huge assortment of air-to-ground weaponry, starting in 1979–1980. However, actually, Project Baz was much more than 'another purchase of new aircraft'; during the related negotiations, Sha'ban went to great lengths to secure the acquisition of the latest electronic warfare systems of French design. Indeed, he went as far as to place an order for a number of systems that as of 1976–1978 existed on paper only, and the development of which became possible only thanks to Iraqi financing. Amongst these were the Thomson-CSF TMV-002 Remora self-protection electronic warfare pod, Thomson-CSF TMV-004 Caiman broad-band offensive electronic warfare pod, Thomson-CSF TMV-018 Syrel electronic intelligence (ELINT) pod, and Matra Sycomor chaff and flare dispenser. Finally, Saddam and Sha'ban prompted the French to develop a custom-tailored version of the Matra/Hawker Siddeley Dynamics AJ.168/AS.37 Martel anti-radar missile. The original Martel was a complex weapon that required three different

A Mirage F.1 of the French Air Force, making a turn to better show the Matra/Hawker Siddeley Dynamics AJ.168/AS.37 Martel anti-radar missile installed under the centreline. The Baz-AR developed for Iraq had the same shape and aerodynamic configuration, but had an entirely new seeker head, making it capable of searching for and targeting US-made air defence systems. (Avions Marcel Dassault via Hugues de Guillebon)

interchangeable seeker-heads to target radars working in three basic bands of frequencies, and was capable of countering Soviet-made air defence systems only. The Iraqis demanded, and the French eventually agreed to deliver, a version that would have a broad-band seeker-head capable of targeting radars working in all three bands, and countering US, British and Israeli-made equipment in service in Israel and Iran. The resulting weapon was nicknamed the Baz-AR ('Falcon, anti-radar').[6]

Unsurprisingly considering the complexity of all the systems in question, the French were to take years to research and develop, and then to deliver everything. For example, of all the weaponry ordered for use by Mirages, only Matra's R.550 Magic was in the process of entering service, while the medium-ranged Super 530F was to follow in 1980. However, once they became available, these systems promised to convert the IrAF into one of the best-equipped air forces in the Middle East. In turn, the Iraqi financing of all these projects was to make the French defence sector capable of not only catching up with the USA in the field of electronic warfare, but even of matching the best available US-made systems.[7]

PROJECT KARI[8]

Negotiations for the sub-project Project Baz-AR in particular, and the resulting availability of the latest high technology of French origin eventually led Sha'ban to the idea of significantly expanding the integrated air defence system that was in the process of emerging as of 1976–1977. The systems acquired from the USSR during the early 1970s – including SA-2s, SA-3s, P-14 radars, and also P-35M and P-37 (ASCC/NACO-codename 'Bar Lock') radars used for wide-area surveillance, early warning and ground control – were all linked up to a single Azurk-1ME ATMS, a family-home-sized first generation computer of Soviet origin, capable of simultaneously calculating intercept vectors for 82 targets. While excellent for coverage of medium and high altitudes, the resulting IADS had massive weaknesses in terms of early warning from low-altitude threats (that is, aircraft underway at less than 300–500 metres/985–1,640 feet altitude). The idea was to expand this system through the addition of French-made TRS.2105 mobile, solid-state, G-Band radars; while having a range of just 70km (43.5 miles), these proved to have an excellent capability for detecting low-flying aircraft. Furthermore, Sha'ban wanted an IADS that would integrate the work of all the electronic warfare assets meanwhile acquired by Iraq, and of all available observation posts, and then expand the capability of the system to monitor developments inside the airspace of all neighbouring countries up to the depth of

The Main Building of the Directorate of Operations IrAF – and the headquarters of the Kari IADS/ATMS – at the former al-Muthanna AB, in the Baghdad suburb of Mansour. (via Ali Tobchi)

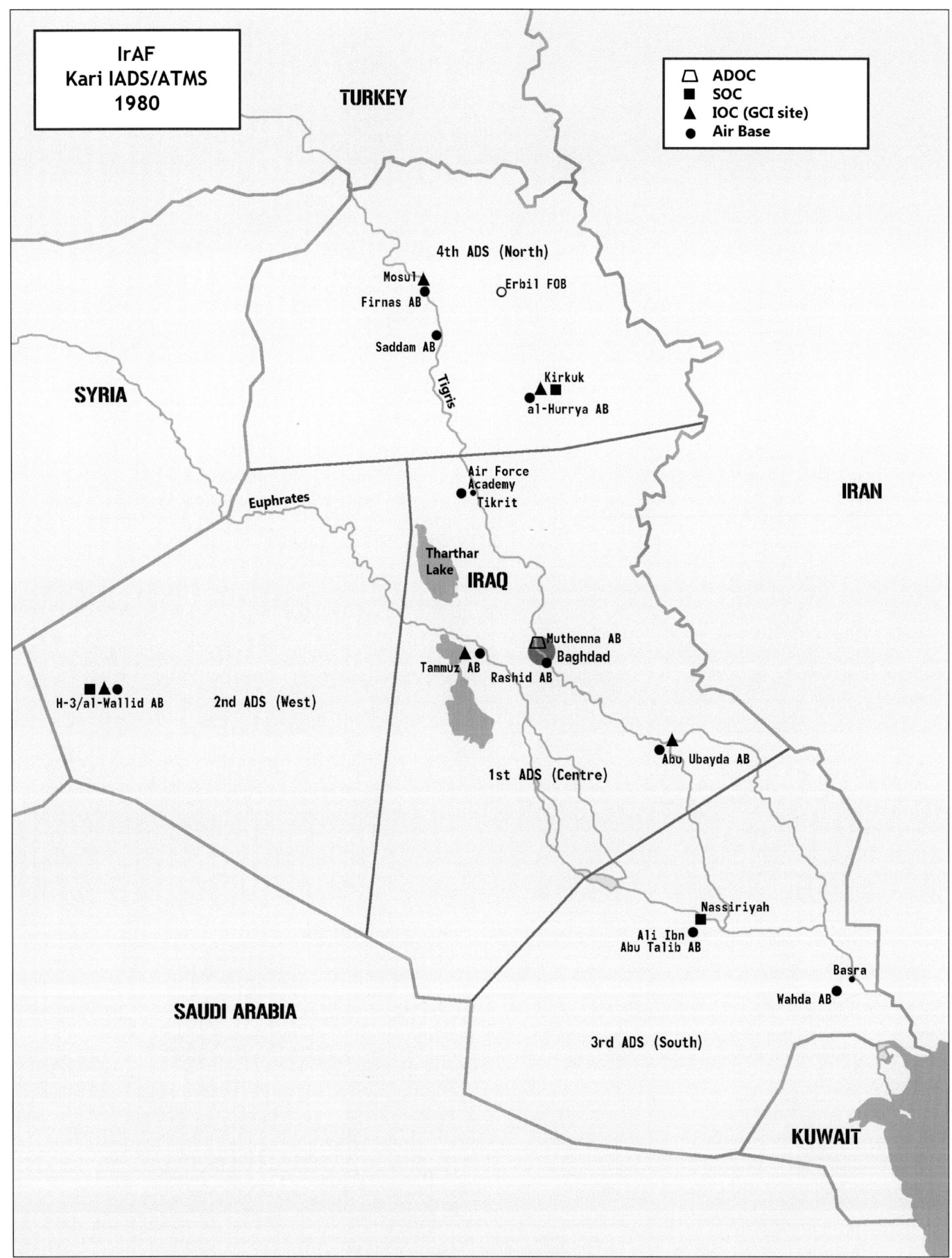

Map of the Kari IADS/ATMS while under construction between 1978–1981. (Map by Tom Cooper)

WINGS OF IRAQ VOLUME 2: THE IRAQI AIR FORCE 1970–1980

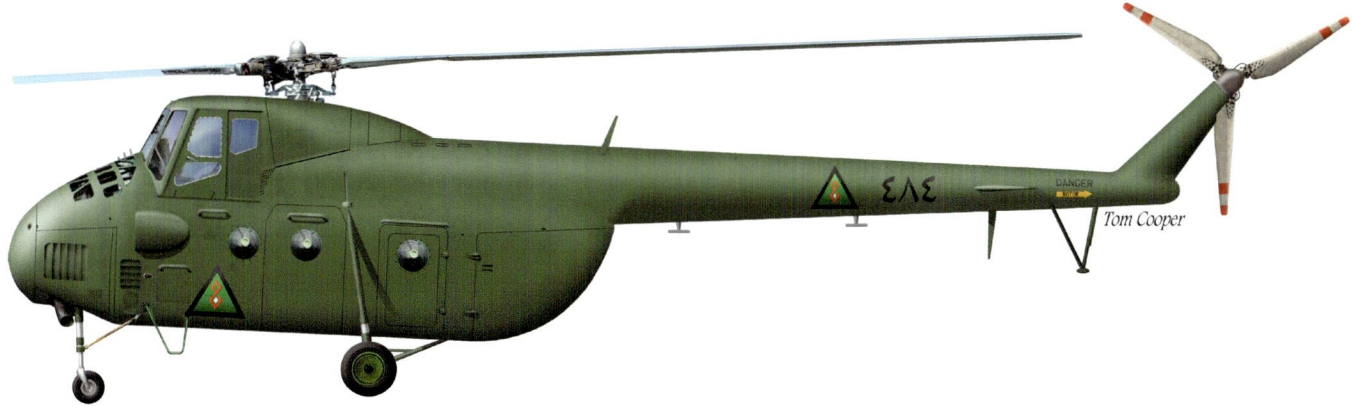

About 35 Mil Mi-4 helicopters were in service with the IrAF as of October 1973, most of them with No. 2 Squadron. The unit flew 83 sorties in support of the hurried redeployment of Iraqi Army units to Syria when its helicopters were used to carry repair teams and spares along the highway from Baghdad to Damascus; these helped with the repair of ground vehicles that suffered malfunctions during the long march. As far as is known, the entire IrAF Mi-4 fleet of the time was painted in dark green overall. Later on, the fleet seems to have received a camouflage pattern including dark sea grey on top of this livery. (Artwork by Tom Cooper)

Out of 12 Westland Wessex HC.Mk 52s acquired by Iraq in 1964–1965, at least nine were still in service with No. 4 Squadron as of October 1973. While originally designed as a troop carrier version capable of carrying up to 16 troops, in Iraq they were mainly used for liaison and VIP transport. All were painted in dark sea grey (BS381C/638) and dark green (BS381C/641) on top surfaces and sides, and an unknown colour on the undersides. Insignia was limited to the fin-flash and a three-digit serial on the boom. The last Iraqi Wessexes were withdrawn from service and replaced by SE.316C Alouette IIIs only in 1979; unlike the Mi-4s, Mi-6s, Mi-8s and Mi-25s, which were handed over to the IrAAC in 1980, the type therefore served only with the IrAF. (Artwork by Tom Cooper)

Iraq was the second biggest export customer for Aero L-29 Delfins in the Middle East. Between February 1968 and April 1974, it acquired a total of 78 jet trainers of this type and had 84 pilots trained on them in Czechoslovakia. All were painted in two layers of clear lacquer mixed with 10 percent and 5 percent aluminium powder, respectively, and wore national insignia in six positions, fin-flashes, and large black serials on the fin (repeated, RAF-style, on the undersides of the wing). The IrAF crest was on the left side of the forward fuselage (and retained there even when many had received camouflage colours in the 1980s), and most early examples had red strakes low along the forward fuselage, and red chevrons on the fin. The survivors were withdrawn from service only in the 1990s. (Artwork by Tom Cooper)

In 1969–1970, Iraq acquired 26 Czechoslovak-manufactured MiG-15bisSB modified as fighter-bombers. The Iraqi Air Force thus became only the second user of single-seat MiG-15s in the Middle East, after the Egyptians. They served as advanced trainers at the Air Force Academy (where Chipmunks, Provosts, and then Zlins were used as elementary trainers, and L-29s as basic jet trainers), but might have seen some action against Barzani's Peshmerga during the Second Iraqi-Kurdish War of 1974–1975. As far as is known, all were left in their original livery as on delivery, including two layers of clear lacquer mixed with 10 percent and 5 percent aluminium powder, respectively, but at least four were re-painted in black and green and served with the acro-jet team of the IrAF. (Artwork by Tom Cooper)

Iraq acquired a total of 20 Percival Jet Provost T.Mk 52s in the 1960s. After serving as advanced trainers at the Academy, they were handed over to the Flight Leaders School, the fighter-weapons school of the IrAF, with which they served well into the 1980s. For the first decade of their service in Iraq, they were left in high-speed silver overall finish, but had large parts of the rear fuselage, top, under, and sides of their wing-tip fuel tanks painted in dayglo orange. Serials were worn RAF-style, on the rear fuselage, and on the undersides of the wing. As well as having two internally-installed 0.303in (7.7mm) Mark 55 machine guns, Jet Provosts could be armed with up to three 127mm unguided rockets per wing, or with three 'banks' of three Sura 68mm unguided rockets per wing. (Artwork by Tom Cooper)

Iraqi Hunters retained their standardised camouflage pattern in dark sea grey (BS381C/638) and dark green (BS381C/641) on top surfaces and the upper half of their fuselage throughout their service: under-surfaces were always left in high-speed silver finish. This jet is shown in standard configuration for combat operations during the October 1973 Arab-Israeli War, and the Second Iraqi-Kurdish War of 1973–1975, including four banks of three 3in (127mm) unguided rockets under each wing (for a total of 24 rockets), a full load of 30mm ammunition for four internal ADEN cannons, and drop tanks. Notable is the application of the serial number on the cover of the front wheel bay. Insets show the crest of No. 29 Squadron, and the way serials were applied on lower wing surfaces. (Artwork by Tom Cooper)

Although taken by surprise by the outbreak of the October 1973 Arab-Israeli War, the IrAF was extremely quick in reacting to it: six units were placed on alert and started deploying to Syria late the same afternoon. The first to go was a mix of MiG-21FLs and MiG-21PFMs operated by No. 9 Squadron; they flew their first combat sorties on the afternoon of 7 October 1973. Because the Iraqi MiG-21s arrived in Syria still lacking camouflage colours (all were painted in 'silver grey', overall, which consisted of two layers of clear lacquer mixed with 10 percent and 5 percent aluminium powder, respectively), the Syrians took care to apply their own camouflage colours to them, and then re-apply their national markings and serials. This is why this MiG-21FL (serial 645) was painted in orange-sand and blue-green on top surfaces and sides, and light blue-grey on undersides. The aircraft survived all the wars until the 1990s, when it was used for test-applications of a radar-reflection-reducing black coating. (Artwork by Tom Cooper)

In memory of their service in Syria, all the surviving MiG-21FLs and MiG-21PFMs of No. 9 Squadron retained the camouflage colours applied by the Syrians during that deployment. This was also the case with this MiG-21PFM, last seen while serving with No. 9 Squadron (crest shown inset) as a point defence interceptor for Saddam and Sahra Air Bases in 1984. Like the MiG-21FL shown above, and following Syrian practice, its serial was repeated high on the fin; a practice unheard of in Iraq. Both of these MiG-21 versions lacked internal guns; their sole armament consisted of a pair of near-useless R-3S missiles. (Artwork by Tom Cooper)

Perhaps the most famous MiG-21MF of the IrAF was this example, which served with No. 11 Squadron from its delivery in 1972, until its withdrawal from service in 1984. Serial number 1019 received one kill marking for an Israeli Mirage claimed during the October 1973 Arab-Israeli War, and then a kill marking for an Iranian F-4E Phantom II claimed on 8 September 1980. Interestingly, although it received this coat of camouflage colours while deployed in Syria, the pattern in question was originally developed by one of the overhaul facilities at Helwan in Egypt and applied on numerous MiG-21PFMs and MiG-21M/MFs in that country – and at least 2 or 3 Iraqi MiG-21MFs (including serial number 1099). This pattern consisted of beige (BS381C/388) and olive drab (BS381C/698) on upper surfaces and sides, and light admiralty grey (BS381C/697) on undersides. Insets show the insignia of No. 11 Squadron and the overhead view of this camouflage pattern. (Artwork by Tom Cooper)

Like MiG-21s and Su-7s, the MiG-17s of No. 7 Squadron, IrAF, also arrived in Syria in October 1973 without camouflage colours and had to be painted before going into battle. The colours used were those standardised by the SyAAF and thus widely available: orange-sand and blue-green on top surfaces, and light admiralty grey (BS381C/697) on under-surfaces. Contrary to Egyptian and Syrian examples, Iraqi MiG-17Fs had no hardpoints under the fuselage, nor had their ground crews to remove drop tanks if any weapons were to be installed on the main underwing hardpoint; they had special attachments for UB-16-57 pods, which were installed directly on the outside of the outboard wing fence. Insets show the crest of No. 7 Squadron and a reconstruction of the camouflage pattern on top surfaces. (Artwork by Tom Cooper)

Sighted in Syria in 1973, and thus widely misidentified as 'Syrian', this was actually one of the MiG-17PFs of No. 7 Squadron, IrAF, camouflaged after its deployment to that country on 9 and 10 October 1973 (Syria had ordered and acquired radar-equipped MiG-17PFs in 1958–1959, but these were taken over by the United Arab Republic Air Force soon after their delivery, redeployed to Egypt and retained there after the dissolution of the union between Egypt and Syria). Indeed, Syrian ground crews went as far as to repeat its serial near the top of the fin. Designated 'MiG-17bis' in Iraqi parlance, the type served as a fighter-bomber in the early 1970s, usually armed with FAB-50, FAB-100, or FAB-250 bombs, or UB-16-57 pods for unguided 57mm rockets, in addition to its two internal 23mm cannons). (Artwork by Tom Cooper)

A reconstruction of the IrAF Su-7BMK serial number 812, as seen in a photograph taken at Hurrya AB in Mosul upon the jet's return from deployment to Syria during the October 1973 Arab-Israeli War. Like most Iraqi combat aircraft of the time, originally the aircraft was painted in 'silver grey' overall (two layers of clear lacquer mixed with 10 percent and 5 percent aluminium powder). Standard Syrian camouflage colours – orange-sand and blue-green, shown here – were applied after its arrival in Syria, during the night of 7 to 8 October 1973. Contrary to Egyptian and Syrian-operated examples, Iraqi Su-7BMKs lacked not only the rear-view mirror atop of the cockpit transparency, but also the second underwing pylon. Their primary armament consisted of up to four FAB-500M-54 bombs. Inset is shown the crest of No. 8 Squadron, IrAF, which had flown Il-28s at earlier times but was re-equipped with Su-7s to act as an OCU. (Artwork by Tom Cooper)

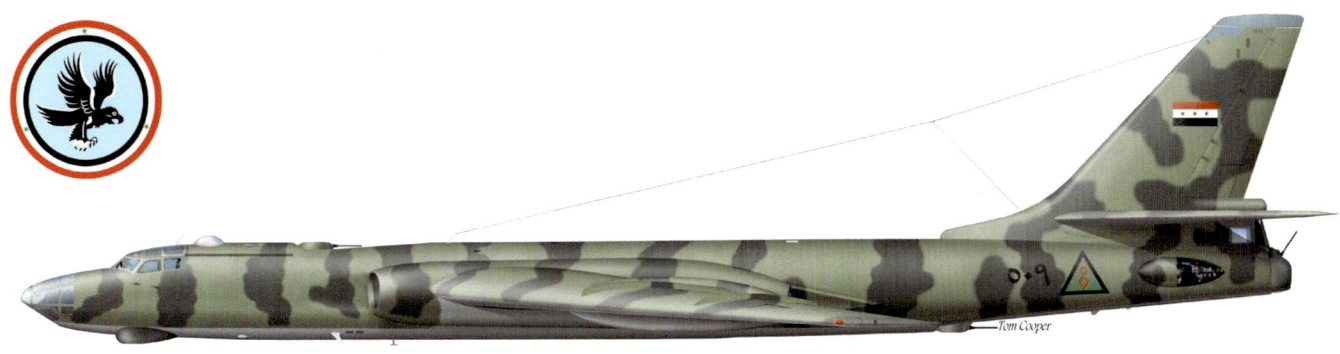

Out of 10 Tu-16s originally acquired by Iraq, only 7 or 8 were still operational as of 1973, when a camouflage pattern was applied to all of them, consisting of the same colours as applied on Tu-22s delivered around the same time. Grey-green (BS381C/283) and dark green (BS381C/641) were applied on top surfaces and sides, while undersides appear to have been left in the original 'silver grey' (two layers of clear lacquer mixed with 10 percent and 5 percent aluminium powder, respectively). 'Roundels' (actually: 'triangles') were applied in the classic positions: on top and bottom wing surfaces and the rear fuselage. The surviving 6 to 7 Iraqi Tu-16s went into the war with Iran still wearing the same camouflage pattern. The inset shows the insignia of No. 10 Squadron, IrAF. (Artwork by Tom Cooper)

IrAF acquired a total of 12 Tu-22 and 2 Tu-22Us, and these were operated by two squadrons: No. 18 received examples with serial numbers 1109 (Tu-22U), 1110, 1111, 1112, 1113, 1114, and 1115; and No. 36 was equipped with serials 1116 (Tu-22U), 1117, 1118, 1119, 1120, 1121 and 1122. All were painted in grey-green (BS381C/283) and dark green (BS381C/641) on top surfaces and sides, and had their under-surfaces in 'silver grey' (two layers of clear lacquer mixed with 10 percent and 5 percent aluminium powder, respectively). The type saw intensive deployment during the Second Iraqi-Kurdish War, and again during the war with Iran. Insets show their most powerful weapons (from left to right): internally carried FAB-9000M-54, FAB-5000M54, FAB-3000M-54 and FAB-1500M-54 bombs. (Artwork by Tom Cooper)

Always insistent on training enough pilots and groundcrews, the IrAF was quick in placing an order for the then brand-new Aero L-39 Albatross training jets from Czechoslovakia almost as soon as these became available for export. The first 50 aircraft had been ordered in 1972, but their deliveries took place only in 1975–76 – and even then, Aero first handed over six L-39Cs originally manufactured for the USSR. The remaining jets were all of the L-39ZO variant, and most wore the Air Force Academy's livery, with large parts of the nose, fin-tip fuel tanks, and fin-tips in red, undersides in medium grey, and a large crest of the Air Force Academy under the front cockpit. Other Iraqi L-39ZOs received a camouflage pattern and served not only as advanced trainers, but also as light strikers armed with unguided rockets. (Artwork by Tom Cooper)

Despite combat attrition, the October 1973 Arab-Israeli War left the IrAF in a diametrically opposite position to all other air forces in the Middle East (including the Israeli): with far more pilots than it had combat aircraft. This and delays in production of MiG-23s resulted in the decision to temporarily re-equip No. 29 Squadron with second-hand MiG-19S pending the delivery of MiG-23BNs that would replace Hunters, its former mounts. This explains why the IrAF operated MiG-19s for a 'second time' after their rather short and unfortunate first tour of duty in 1959–1964. Available photographs indicate that all the aircraft were painted in mid-grey overall, and wore the standard set of national markings and serials. The insignia of No. 29 Squadron, as used until the unit was re-equipped with MiG-23BNs in 1976–1978, is shown inset. (Artwork by Tom Cooper)

In 1976, the IrAF acquired its first dedicated reconnaissance aircraft in more than a decade: four MiG-21Rs. Operated by a dedicated flight of No. 70 Squadron, these arrived painted in two layers of clear lacquer mixed with 10 percent and 5 percent aluminium powder, respectively. More importantly, they could be equipped with D and P-pods, of which the former were of special interest for the Iraqis, because they enabled them to monitor and record the work of Iranian early warning radar sites and other, radar-supported, ground-based air defences. The serials of the original four examples remain unknown: serial 1828, illustrated here (with the D-pod under the centreline and the P-pod in the lower left corner) was probably one of the extra examples acquired as attrition replacement. Operations along the Iranian border always required the carriage of 400-litre drop tanks on outboard underwing pylons. (Artwork by Tom Cooper)

Due to simultaneous demands from Algeria, Egypt, Syria and North Vietnam, deliveries of MiG-21M/MFs ordered by Iraq in 1970 lasted well into 1974–1975. The last batches to reach the IrAF thus arrived already wearing a new, standardised camouflage pattern in beige (BS381C/388) and olive drab (BS381C/298) on top surfaces and sides, and light admiralty grey (BS381C/697) on undersides. Furthermore, they were delivered together with new Egyptian-designed 800-litre drop tanks (carried under the centreline only), and new and much improved R-13M infra-red homing, air-to-air missiles (one of which is shown on the inboard underwing pylon). This MiG-21MF was operated by one of two Rashid-based units, either No. 11 or No. 70 Squadron, from the mid-1970s well into the 1980s. (Artwork by Tom Cooper)

WINGS OF IRAQ VOLUME 2: THE IRAQI AIR FORCE 1970–1980

The last few Su-7BMKs delivered to the IrAF in 1973 and in time to be deployed to Syria arrived already wearing the same camouflage pattern as standardised for MiG-21MFs manufactured in the mid-1970s: consisting of beige (BS381C/388) and olive drab (BS381C/298) on top surfaces and sides, and light admiralty grey (BS381C/697) on undersides. Unlike Egyptian and Syrian Su-7BMKs, Iraqi Su-7BMKs seem to have never received a rear-view mirror installed atop their cockpit transparencies, and they had only one underwing hardpoint under each wing. This example is shown together with the crest of No. 5 Squadron, which was heavily involved in the Second Iraqi-Kurdish War of 1974-1975. (Artwork by Tom Cooper)

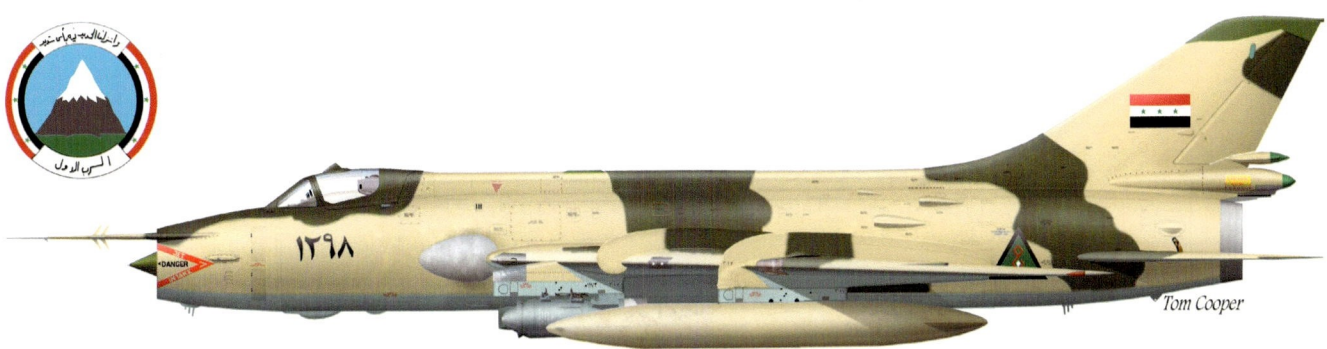

Midway through the October 1973 Arab-Israeli War, and without any kind of an order from Baghdad, Moscow started delivering Su-20s to Iraq. After thorough testing, the IrAF accepted the new type with great enthusiasm and rushed it into service with No. 1 Squadron (crest shown inset) as soon as the unit was back from deployment to Syria. The aircraft arrived already wearing a further development of the camouflage pattern applied on MiG-21MFs and Su-7BMKs, consisting of beige (BS381C/388) and olive drab (BS381C/298) on top surfaces and sides, and light admiralty grey (BS381C/697) on undersides, and wore serials 1162–1173 and 1295–1315. The entire fleet was grounded in September 1980, and thus No. 1 Squadron was sorely missed during the opening Iraqi strike on Iran. (Artwork by Tom Cooper)

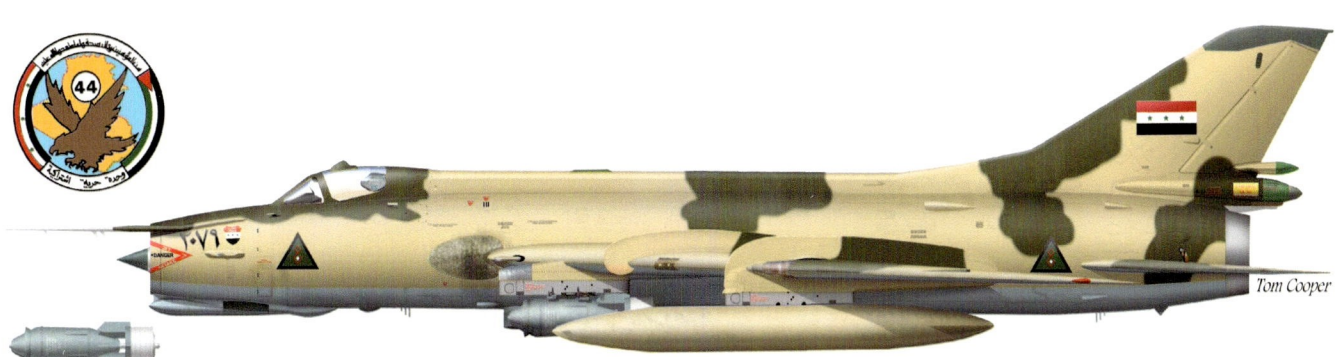

In response to Iraqi demands, the Soviets developed the Su-22, a variant including several major improvements, like a lengthened fuselage with narrower intake, the DISS-7 Doppler speed and drift angle system under the intake, a Fon laser rangefinder and the more powerful R-29 engine. The first batches (serials in ranges 1700, 1800, and 20xx) were painted in the same camouflage pattern as Su-20s before them, including beige (BS381C/388) and olive drab (BS381C/298) on top surfaces and sides, and light admiralty grey (BS381C/697) on undersides. They received additional national markings on the forward fuselage. In 1980, No. 44 Squadron was granted permission to apply a marking in the form of a shield with Iraqi national colours and an inscription ('Iraq' in English and Arabic) for its excellent performance. This jet from that unit is shown configured with FAB-500M-54 bombs, as used to crater the runways of Iranian air bases during the air strikes of 22 September 1980. (Artwork by Tom Cooper)

vii

With the production of MiG-21bis only starting in 1977, and the first export examples being 'reserved' for Ethiopia, the Soviets were critically short of airframes that they could deliver to Iraq in response to the order from 1978. Therefore, they used older MiG-21SM and MiG-21SMT airframes; these were overhauled, received new, large intakes, different spin and fin, and more powerful engines and were delivered to Iraq as MiG-21bis, starting in 1979, wearing a modified camouflage pattern originally developed for the MiG-21MF-series. The first batch received serials around 2102, and went to the newly established No. 47 Squadron, the insignia of which is shown in the inset. They arrived together with additional R-13M missiles (shown on the outboard underwing pylon) and brand-new R-60MKs (inboard underwing pylon). (Artwork by Tom Cooper)

Probably the greatest disappointment for the IrAF of the 1970s was MiG-23MS. Much anticipated in Iraq, this early variant suffered from poor manufacturing quality, poor flight-testing (resulting in poor quality of flight manuals), poor navigation-attack suite and poor armament. It took No. 39 Squadron (insignia shown inset) nearly four years to make it fully operational and rewire surviving mounts for the carriage of R-13M missiles (shown on underwing hardpoints). All Iraqi MiG-23MS wore the same standardised camouflage pattern consisting of beige (BS381C/388), dark brown (BS381C/411) and olive drab (BS381C/298) on top surfaces and sides, and light admiralty grey (BS381C/697) on undersides. No. 39 Squadron was one of very few IrAF units granted permission to apply a unit crest on its aircraft. (Artwork by Tom Cooper)

MiG-23BNs proved another disappointment, not only because of their poor manufacturing quality, but also because of not being equipped to the standard originally ordered and paid for by Iraq. Having no other choice, the IrAF eventually equipped two units with them: No. 29 Squadron (the new crest, introduced in 1976, shown inset), and No. 49 Squadron. All wore the same camouflage pattern as MiG-23MS in beige, dark brown and olive drab on upper surfaces and sides, even though there were significant differences from aircraft to aircraft. Undersides were in light admiralty grey: the Soviet practice of painting the lower rear fuselage in light or medium grey was quickly discontinued in Iraq. This jet (serial 1428) is shown as configured for the opening airstrikes against Iranian air bases on 22 September 1980, with a parachute-retarded FAB-500M-54 bomb on the under-fuselage hardpoint. (Artwork by Tom Cooper)

With relations with the USSR troubled due to Iraq's politics vis-à-vis the Kurds and the Iraqi Communist Party, from the mid-1960s Baghdad increasingly turned to Czechoslovakia for arms and advice. After buying spares for MiG-15s and MiG-17s, in 1967 the first order for 20 Aero L-29 Delfins and the training of pilot instructors and ground personnel was placed. Twenty L-29s followed in 1968 and another 38 by April 1974. One of them was this example with serial 1124, shown wearing the gaudy markings of the Air Force Academy. (via Ali Tobchi)

In the course of negotiations for the Iraqi-Soviet Friendship Treaty, Baghdad demanded supersonic Tupolev Tu-22 bombers: 14 were delivered by October 1973 but arrived without any weapons. Amongst them were two Tu-22U conversion trainers. This photograph shows the second, serial 1116, operated by No. 36 Squadron, inside one of the revetments at Taqaddum AB. (via Ali Tobchi)

The acquisition of Il-76MDs in 1978-1980 significantly improved the strategic mobility of the entire air force; henceforth, entire squadrons could be redeployed to western Iraq, Syria or Egypt at short notice. The type proved highly popular in service, and although most wore the colours of Iraqi Airways, they were operated by the IrAF, which eventually acquired more than 40. (Milpix/Martin Hornliman Collection)

Unit Insignia of RIrAF & IrAF Squadrons
1931-2003

No. 36 Squadron
1973-1990

No. 37 Squadron
1979-1995

No. 39 Squadron
1974-1995

No. 44 Squadron
1977-1992

No. 47 Squadron
1979-1995

No. 49 Squadron
1974-1992

No. 59 Squadron
1974-1992

No. 79 Squadron
1981-2003

No. 81 Squadron
1984-1991

Project 84
(service introduction of
MiG-25, 1979-1981)

No. 93 Squadron
1986-1995

No. 96 Squadron
1984-2003

No. 109 Squadron
1987-2003

No. 115 Squadron
1988-1991

Flight Leaders School
1970-1991

(depending on the local terrain) 100–150 kilometres (62–93 miles). Resulting negotiations with DIA led to the contract for Project Kari: the construction of a semi-automatic command, control and communications system (C3) largely using components (mostly computers) made by Thomson-CSF, electro-optical cables (made by the Swedish company Ericsson), and encrypted communication systems (made by the Swiss company Crypto AG).[9]

The centrepiece of Project Kari was to become the Air Defence Operations Centre (ADOC) – a massive, hardened facility constructed at the former Muthenna AB in Baghdad. While in direct control over electronic warfare units, the Kari ADOC was to be in permanent communication with – and thus in control of – four Sector Operation Centres (SOCs, each inside its own building, all hardened and largely constructed underground), each of which used French-upgraded, Soviet-made Bastion-3E(F) systems as command posts to exercise operational control over whatever manned-interceptor, SAM, and anti-aircraft artillery units, radar stations and ground observation posts were deployed within its geographic area of responsibility.[10] With the help of the Kari communications network, the radar picture from every early warning radar station in Iraq was fed not only to the three-to-five ground control intercept stations assigned to each of the SOCs, but also into the ADOC, and the other way around; to every individual SAM site and anti-aircraft artillery battery. In this fashion, the Air Chief – or, more often, his Deputy Operations – in Baghdad was to become capable of exercising direct command and control of all IrAF operations from one command post. Moreover, and contrary to the situation in Egypt and Syria in October 1973, the ADOC was connected to the headquarters of all corps-sized formations of the army, which was as important because the latter operated not only a large number of anti-aircraft artillery pieces of 23mm and 57mm calibre (including the mass of self-propelled and radar-controlled ZSU-23-4 Shilka and ZSU-57-2s), but also all the available SA-6, SA-8, SA-9 and SA-13 SAM systems. To run the Kari IADS, the IrAF established the Air Defence Command. This assumed control over all available radar stations and the five missile brigades (which included a mix of 14 SA-2 and 21 SA-3 SAM sites) deployed for the protection of major air bases and crucial industrial facilities, plus a large number of anti-aircraft artillery units of the IrAF, meanwhile re-equipped with 12.7mm and 14.5mm heavy machine guns, and 23mm, 37mm and 57mm automatic guns.[11]

Initiated in 1978, the construction of Kari was be completed in 1982; in the meantime, the Air Defence Command IrAF was organised into four Air Defence Sectors (ADS), as follows:

- 1st ADS (Centre): SOC at-Taji
- 2nd ADS (West): SOC al-Wallid AB
- 3rd ADS (South): SOC an-Nasiriyah
- 4th ADS (North): SOC Kirkuk

TRAPEZOIDS[12]

Contracts for Projects Baz, Baz-AR and Kari were still not all. On the contrary, the French agreement to deliver fighter-bombers and advanced weaponry was conditional on the Iraqi agreement to let a group of Belgian and French companies take over and complete the construction of Saddam AB, which was planned to serve as the main hub for F.1 operations in Iraq. Officially named Saddam Air Base, and having a design based on that of Abu Ubayda Ibn al-Jarrah (which was subsequently developed into the second hub for Mirage operations in Iraq), the new facility included only one runway, but enough hardened aircraft shelters (33) and support facilities for three squadrons of F.1s, and a search and rescue (SAR) flight. Its construction began in 1978 and was completed in 1982.

Meanwhile, Sha'ban's work on developing the IrAF following Riyadh's ideas, and good cooperation with Belgrade, resulted in the third major contract for the Yugoslavs, signed in 1978 and stipulating the construction of one major new air base and numerous additional – but smaller – facilities, all based on the trapezoid form. The biggest – and most heavily hardened – of these became Project 1100, or Qadisiyah AB (named in reference to the Battle of Qadisiyyah in the year 636). It was positioned in north-western Iraq, halfway between Baghdad and the nearest point on the border with Syria. Designed and constructed in cooperation between the Federal Directorate of Supply and Procurement in Belgrade, and the companies Aeroengineering (Belgrade), Vranica (Sarajevo), and Granit (Skopje), between 1981 and 1987, it received not only two runways, but also more than 30 hardened aircraft shelters, each with an entry at the front and the rear, its own taxiway that connected it directly to the nearest runway, and extensive training and housing facilities. During the same period, numerous other 'trapezoids' – all comparable to the original 202A in form and size – were constructed near Dhuluiya and Tuz Churmatu (in place of the old airfields from the 1910s), and then near the water-sources of Talha (better known as 'Mudaysis' in the West) and

A rare photograph from the early 1980s, showing a group of IrAF ground controllers in front of one of the consoles of the Kari IADS/ATMS at the Air Defence Operations Centre at al-Muthanna AB. (Albert Grandolini Collection)

This close-up of an aerial photograph of Samarah East, one of the 'satellite' airfields of the Iraqi Air Force Academy, shows a typical Yugoslav-constructed 'trapezoid', a design based on that of Project 202B, where every hardened aircraft shelter had its own taxiway connecting it to the end of the runway. An enemy attacking such a facility had to destroy every single taxiway and the runway in order to render the base non-operational. (Photo by Martin Rosenkranz)

Memorial at the entry to Qadessiya AB, recording the Yugoslav companies involved in this massive project. (Photo by Igor Dujmusic)

Jalibah, in south-western Iraq and Qualat Sukar and Amarah in southern Iraq.

NEW DOCTRINE AND STRATEGY[13]

Combined with the acquisition of new high technologies from France (and elsewhere in Europe) was the emergence of new doctrine and strategy for the IrAF. The process in question began with a series of conferences held in 1975 and 1976, during which veteran officers of the October 1973 War and the Second Iraqi-Kurdish War were encouraged to discuss their experiences. One of the resulting conclusions was that the Iraqi armed services should train 'as hard as the Israelis'. Correspondingly, the GMID received the task of obtaining as many training manuals of the Israeli Defense Force as possible. However, once these became available and were translated (to Arabic and also to English), in 1977–1978, the army-dominated General Command in Baghdad had second thoughts; it concluded that the available equipment was ill-suited for such intensive training and was likely to wear out much too quickly, making the resulting training too expensive. Eventually, the solution was found by selecting only one unit – the 3rd Armoured Division (colloquially known as 'Salahuddin/Saladin Force' in Iraq), which was considered the army's best – to be trained according to the resulting new training manual, the first document of this kind completely developed in Iraq. Sha'ban proved much less reserved; instead of training just one unit of the air force, he decided to reorganise the training of his entire force.

As a result, by 1978, by when the air force was expanded to about 25,000 personnel, the High Command IrAF had four major departments:

- Directorate of Air Operations: essentially a planning cell for all flying operations, commanded by the (first) Deputy Chief of Staff, also titled 'Deputy Operations'
- Directorate of Air Training: a new planning cell for both the development of new doctrine, strategy, and tactics, and for training replacements according to new ideas commanded by the (second) Deputy Chief of Staff, titled 'Deputy Training'
- Directorate of Air Defence: a planning cell for all air defence operations, led by the (third) Deputy Chief of Staff, titled the Deputy Air Defence
- Directorate of Logistics and Planning: a cell responsible for acquisition of replacement and new equipment, ammunition and other supplies, led by the (fourth) Deputy Chief of Staff, titled the Deputy Logistics

This reorganisation at the top resulted in a reorganisation at mid-level, some of which was certainly motivated by Saddam's obsession with security. The most notable change was the reorganisation of wing-level units, which was henceforth closely reminiscent of the US Navy; correspondingly, and for logistical and training purposes,

MIDSUMMER NIGHT'S INTERCEPTION

In summer 1978, in reaction to several violations of the airspace over western Iraq, No. 39 Squadron was redeployed from Taqaddum to Wallid AB. At an unknown date, late in the evening, it was the turn of Captain Hisham Ismail Barbouti to stand alert:

> Alarm bell[s] scrambled me and my Number 2 into aircraft. We powered up engines and requested the ground control to turn on the runway lights. They replied that this would take time until the ground crew had powered up the generator. Having no time to waste, we used the lights of our aircraft to find the way to the runway, engaged the afterburner and took off. Climbing to an altitude of 8,000 metres, I continued accelerating while warming up my radar and requesting an intercept vector. The ground control ordered me in a western direction and advised about the altitude of the target. When I powered up the radar it detected two targets. I adjusted the course, locked on and requested permission to open fire. The ground control replied with 'hold on!'. I couldn't because I was at Mach 1.4 and rapidly approaching! The controller then granted permission to fire, and I released one missile. Moments later, the controller ordered me into a right turn, because my Number 2 was approaching with an even higher speed from behind. As I was braking, I saw an explosion in the direction of the enemy. We returned to base, landed safely, and walked back to the squadron ready room. The word there was that the Air Chief was about to arrive from Baghdad…. Everybody gathered in the conference room, including the base commander, our squadron CO and all pilots from my unit, my ground controller and several air defence officers, and a Soviet expert who studied the radar tape. We discussed all the details of my mission, step by step, and the conclusion was that around the time my missile detonated, one of [the] enemy targets disappeared from the screens of our ground radars. As in confirmation, our ground observers saw a burning aircraft crashing to the ground to the west of the border with Jordan. However, the base commander and our squadron CO were not convinced. Although the next day I received a telephone call from the Minister of Defence, and he congratulated me for my feat, this flight was soon forgotten by all of [the] superior officers. I do feel our Israeli enemy has received a painful blow, though, and remain very proud about that mission.'[14]

Israel never published any kind of commentary in response to this Iraqi claim, and it remains unclear if the Iraqi pilot had hit and shot down or merely damaged his target – which he clearly identified as an F-15 – or if his missile detonated prematurely. Nevertheless, Hisham Ismail Barbouti was subsequently decorated, advanced in rank, and appointed the CO of the Su-22-equipped No. 109 Squadron.

A MiG-23MS from No. 39 Squadron, seen on the finals to landing, armed with R-13M missiles on underwing hardpoints, and R-3S missiles on under-fuselage stations. Notable is the centreline drop tank and the relatively 'awkward' position of the national insignia on the wing: these were applied when the aircraft was parked on the ground, with wings fully swept back. (Tom Cooper Collection)

squadrons were assigned to wings depending on the type of aircraft they flew, but operational control of them was exercised by base commands. Now designed to closely cooperate with the Directorate of Air Operations, the newly established Directorate of Air Training was a large organisation controlling not only the Air Force Academy, but also about a dozen OCUs and the FLS. The latter was re-equipped with surviving Hunters (in addition to the Jet Provosts it was still flying) and became not only responsible for the operational training of future flight and squadron commanders, but an 'aggressor' asset, capable of mimicking operations by enemy air forces – like those of Iran and Israel. At the same time, and for operational purposes, the FLS was directly subordinated to the High Command.

6

THE RISE OF SADDAM

Except for the development of Project Baz, there are next to no historical records about the Iraqi Air Force in 1978–1980. One reason is that its further growth was overshadowed by the politics, particularly Saddam Hussein's final climb to power; another was due to the events immediately after. Therefore, much of what happened with the IrAF during this period can only be deduced on the basis of what is known about a few directly related affairs, and a mass of indirectly related ones.

MOSCOW'S SALE[1]

The start of events decisive for the way in which the expansion of the IrAF was pursued in 1978–1980 can be traced back to 1976, when Saddam Hussein – soon after placing orders for French arms and two nuclear reactors – decided to distance members of the CIP from the RCC and all other positions of authority. To say that the Soviet leader Leonid Ilyich Brezhnev and the Politburo in Kremlin were outraged would be an understatement; by the time they found out that the contracts between Baghdad and Paris included Thales providing maintenance services for Soviet military equipment, they were infuriated. Here it is important to keep in mind that during the first half of the 1970s the USSR had reached the peak of its economic and military power; for several years at least, its economy was growing faster than that of the USA and Western Europe, and it equalised Western advantages in nuclear weapons. From the standpoint of both the Kremlin and the General Staff in Moscow, there was nothing matching the quality of Soviet arms. However, what the aged apparatchiks had failed to understand was that within only a few years the situation of their country would change dramatically; not only that the Soviet technology would begin lagging behind the Western, but pervasive structural problems caused by a phenomena known as the 'undynamic gerontocracy' – the sheer age of the Soviet leadership – would then result in economic stagnation and deep crisis. To the leadership in the Kremlin, this was entirely unexplainable. Instead of adapting to the new situation and initiating reforms, it reacted with ever less patience and understanding. Unsurprisingly, nobody in the Politburo could offer a clear-cut explanation for why, in spite of over a decade of arms deliveries to Iraq, this could not buy them a much greater amount of influence in Baghdad.

The leadership in Moscow was exercising immense pressure to dissuade Bakr and Saddam from concluding deals with France while negotiations were still going on and it even reinforced its efforts once contracts were already signed. Concerned about possible repercussions (one should keep in mind that around the same time the USSR imposed an arms embargo upon Syria, 'just because' it dared openly confront the Palestinian Liberation Organisation, which was supported by the USSR), in July 1976, President Bakr sent his foreign minister Tariq Aziz to Moscow to open negotiations for a new arms deal with the Soviets and announced additional investment in the economic development of northern Iraq.[2]

Encouraged, Brezhnev and the Politburo reacted with 'carrot and stick' tactics; they offered additional MiG-23s at concessionary prices while threatening to call in Iraq's debt, and to impose an arms embargo should Baghdad continue purchasing French weapons. Emboldened by good cooperation with France, Bakr and Saddam remained unimpressed; on the contrary, they began demanding the latest Soviet combat aircraft and electronic warfare systems, while the latter launched a campaign of persecution of the ICP, arresting and executing dozens, and driving hundreds of Iraqi communists out of the country. By the end of 1976, Moscow found itself forced to offer political asylum to some of the survivors. Surprisingly enough, the Politburo did not sanction Baghdad. Instead, it became embroiled in an exchange of ever-fiercer critique; Moscow complained about the mishandling of the ICP, and Baghdad's harsh pacification of the Kurds, it expressed readiness to accept Israel's right to existing within its pre-1967 borders, and it criticised Iraq's support for splinter groups of Palestinian militants at odds with the PLO. However, Saddam was not to be outdone; he voiced opposition to the Soviet diplomatic and military support for Ethiopia during the Ogaden War of 1977–1978 and began providing aid to the Eritrean secessionist rebels and to 'Arab' Somalia. For all practical purposes, relations between Iraq and the USSR were worsening, and negotiations for the new arms deal went nowhere until September 1977, when, amid rumours about Bakr's poor health, Saddam assumed control of all aspects of Iraq's oil policy, and then convinced the president to relinquish control of the Ministry of Defence to his cousin (and Bakr's son-in-law), Adnan Khayrallah at-Talfah; although a civilian, the latter was even promoted to the rank of Major General. Khayrallah promptly initiated a rapid promotion of selected favourites in the armed forces, all of whom supported Saddam Hussein's way of dealing with Moscow. Saddam's men

Saddam Hussein and Ahmed Hassan al-Bakr, seen in 1978. (Government of Iraq)

Major General Mohammed Jassim Hanish al-Jaboury, Air Chief from June 1978 until June 1984. (via Ali Tobchi)

Saddam Hussein, seen shortly after assuming power in Baghdad in July 1979. (Government of Iraq)

knew no diplomacy vis-à-vis the USSR; when, during a meeting related to the new arms deal, Soviet representatives complained about Baghdad's practice of awarding lucrative contracts to Western companies, one of the Iraqis snapped back 'we have the money and so we can afford to buy the best.'

Once again, it was Bakr who found a solution; in spring 1978, he effected an early retirement of Major General Hamid Sha'ban – a 'Saddam's man' – and replaced him with 'Soviet-friendly' Major General Mohammed Jassim Hannish al-Jaboury as the new Air Chief. Semi-satisfied, Moscow agreed to meet Iraqi demands. The result was the biggest Iraqi-Soviet arms deal of all time. Worth US$3 billion, and signed in September 1978, amongst other things this stipulated the delivery of Post-2MK and Post-3M (ASCC/NATO-codename 'Swing Box') ELINT-systems (capable of detecting the emissions of airborne radars like the AWG-9 of Iranian F-14s), SPB-7 and SPO-8 ground-based air defence jammers (ASCC/NATO-codename 'Tub Brick' and 'King Pin', respectively, capable of disturbing the work of airborne radars), and R-834 and R-834b systems for jamming ultra-high frequency (UHF) communications, as used by the Iranians. Moreover, the IrAF was granted permission to order electronic warfare systems for its Tu-16 and Tu-22 bombers, and a total of 18 Ilyushin Il-76MD transports, 54 MiG-21bis, 54 MiG-23MF and six MiG-23UB, 18 MiG-25PDS, 18 MiG-25RBs, eight MiG-25PUs and 18 Su-22Ms, the first of which were to arrive in early 1980.[3]

MIDSUMMER COUP[4]

Lengthy and troublesome negotiations with the Soviets left their traces upon relations between Bakr and Saddam, too. In late 1978, the president launched a strong initiative for an Iraqi-Syrian union. Contemplated and negotiated time and again since 1963, the idea prompted Damascus (which felt dangerously exposed to renewed Israeli aggression because of the decision of Egyptian President Anwar el-Sadat to enter peace negotiations) to re-open its border with Iraq and the T-Pipeline, and Bakr then financed the second Syrian acquisition of MiG-23BN fighter-bombers from Moscow.[5]

However, Saddam could not care less about Soviet feelings, and had his own ideas about the future of Iraq. Keen to prevent Hafez al-Assad from assuming the position of a new pan-Arab leader and the champion of support for Palestinian resistance movements, after concluding the destruction of the ICP and thus reinforcing his position within the Ba'ath, he prepared his final ascent to the top position. In April 1979, during the visit of the French Prime Minister Raymond Barre to Baghdad, another arms deal worth US$1.5 billion was signed, and Iraq obtained a guarantee for the construction of the two reactors in Tuwaitha, together with associated deliveries of uranium. Only a month later, Saddam concluded an arms deal with Spain; worth about US$1 billion, this was primarily related to the acquisition of general-purpose bombs for Mirages. Finally, during a secret meeting of the RCC on 11 July 1979, Khayrallah and his son Adnan began urging Bakr to resign. Eventually, the president succumbed to pressure and on 16 July 1979, he appeared on television and demanded to be relieved of all positions for 'health reasons'. Immediately afterwards, Saddam launched a major purge of potential internal opposition; he announced a 'coup conspiracy… supported by Syria', and involving top members of the RCC, foremost the Secretary General of the Ba'ath Party, Mohyi Abdel Hussein al-Mashhadi. The latter was arrested on 28 July, and his confession extracted under torture implicated more than 80 top officials of the party, including the Minister of State for Kurdish Affairs, Khaled Abdel Osman, and the Deputy Prime Minister Adnan Hussein al-Hamadani. Up to 60 of the 'conspirators' – including five members of the RCC – were executed during the following days, several by Saddam himself.[6]

This 'midsummer coup' marked not only the end of the unification process of Iraq and Syria, but also the moment Saddam – with a single blow – took over as the Chairman of the RCC and thus the President of Iraq and the Commander in Chief of the Armed Forces. To secure the loyalty of the latter, only a month later, he nearly doubled the salaries of all military officers. For all practical reasons, he was now subsidising the officer corps, much of which already owed its positions to him, and was keeping the armed forces happy not only with the help of lavish spending on new equipment, he was also keeping them in the tight grip of his patronage networks and intelligence services. From that point onwards, there was nobody in Iraq left in a position to openly challenge Saddam, and his and the fate of the country – and that of the IrAF – became in effect the same.

GOLDEN TIMES

At first glance, it is all too easy to conclude that only months after Jaboury assumed the command of the Iraqi Air Force, his service was back to the practice of ordering large numbers of combat

(From left to right) Lieutenant General Muhammad Jassam Hanish al-Jaboury, Air Chief, 1978–1984; al-Hakam Hassan Ali at-Tikrity, and Colonel Salim Sultan al-Basu. The latter served as an Su-7 pilot, then Squadron CO, and then as CO Firnas AB, before being reassigned to the position of the Director Aerial Movements, a sub-position of the Directorate of Air Operations, in 1978. (via Ali Tobchi)

Table 3: Commanders of the IrAF, 1970–1980		
Period	Name	Background & Notes
18 Jul 1968 – July 1970	Hardan Abd al-Rhefal at-Tikriti	second tenure; dismissed and forced into exile in 1970
Jul 1970 – 1 Jun 1973	Hussein Hiyawi Hamash at-Tikriti	
1 Jun 1973 – 1 Jun 1976	Nima'a ad-Dulaymi	pilot
1 Jun 1976 – 1 Jun 1978	Hamid Sha'aban at-Tikriti	first tenure; pilot
1 Jun 1978 – Apr 1984	Mohamed Jassim Hanish al-Jaboury	pilot

aircraft from the USSR. Indeed, while some of the IrAF veterans recall the next two years as the 'golden times', others commented on 'the Return of the Sukhoi Mafia'.[7] However, not only had related negotiations been going on for years before, but the actual reasons for this 'reversal' were far more complex than they might appear at first glance.

Within the IrAF, there was still a widespread belief that the Soviets were capable of delivering plenty of aircraft at short notice, and also that their jets were robust, resilient to combat damage and easy to maintain. Far from being an indication of 'preference for quantity over quality', this was a belief based on actual experience, some of which went so far that when studying the Mirage F.1 they were about to acquire, many of the air force's officers were wondering if the aircraft could ever survive any kind of combat damage at all, not to mention the kind of hits survived by Su-7s and Su-20s during the two wars of 1973 and 1974–1975. Furthermore, and just like so many Western experts, the Iraqis remained deeply impressed by the exceptionally high attrition of their air force during the last war with Israel; they needed an attrition reserve, just like they wanted to expand the capacity and capability of their air force.

At the political level, and despite appearances, Saddam agreed to secure relations with Moscow out of his growing concerns vis-à-vis Iran. The IIAF had received its first F-14A Tomcats in January 1976, and received the last few in early 1978, after which the Iranians demonstratively ran several spectacular test-firings of their AIM-54A Phoenix long-range missiles. Not only were the Tomcats with their superior weapons systems leaving deep impressions upon many in Baghdad (and especially Saddam Hussein), but the Iraqis were at least as impressed with Tehran acquiring not only the aircraft and weaponry, but also the complete infrastructure necessary to maintain them at home. Saddam and his aides wanted a similar deal for Iraq, and France agreed to provide exactly that. In the meantime, they had to keep the IrAF satisfied by bridging the time until the IrAF could catch up with the Iranian advantage in technology. That is why they eventually agreed to buy additional aircraft from the USSR.

Exactly what role Major General Jaboury played in all these 'games' remains unknown, because very little is known about this officer, except that he came from a clan with a long history of affiliation with the IrAF, and what has been mentioned above about his service in Syria in October 1973. Some veteran air force officers describe him as a popular commander, widely appreciated by the rank and file. What is also known is that Jaboury brought with him at least three new deputies: Brigadier General Ahmad Kahyri was appointed the Deputy Operations; Brigadier General Salim Sultan al-Basu the Deputy Air Defence; and Brigadier General Osama al-Yawer the Deputy Logistics and Planning. As far as is known, all had the background of flying Soviet-made fighter-bombers at an earlier time. For example, Basu first served on Hunters, but then converted to Su-7s and commanded not only a squadron of these, but also Firnas AB until June 1978.[8]

PLAYING WITH FIRE

The reasons for Saddam's concerns about Iran became obvious in early 1979. Following months of continuously spreading mass protests, in February that year Shah Reza Pahlavi left the country, and a new government led by Grand Ayatollah Sayyid Ruhollah Moussavi Khomeini took over. Initially at least, Baghdad sought an accommodation with the new authorities in Tehran. However, Khomeini reacted by announcing his intention to 'export' his quasi-Islamic Revolution elsewhere in the Middle East. In turn, this brought the Iraqi Shi'a gravitating around the Islamic Dawa Party to the idea; they began calling on their leader, Muhammad Baqir as-Sadr, to become the 'Iraqi Ayatollah Khomeini' and lead a revolt against Saddam. Popular protests erupted in Iraq in May 1979, rapidly spreading from Baghdad into the south. Saddam Hussein reacted in an unexpected fashion; instead of promptly deploying the armed forces and security services, he appeared to wait for several days. Actually, Iraq's oil-wealth had enabled him to equip his multiple parallel secret services with sophisticated means of surveillance and control; indeed, during the mid-1970s, units equipped for communication control grew faster in terms of equipment and manpower than all the branches of the armed forces combined. Trained by the French and Japanese, their personnel achieved an exceptionally high standard of proficiency. Combined with the expansion of the country-wide telecommunication system, this meant that Saddam's security services were now in fact capable of monitoring every single telephone call. Unsurprisingly, they quickly penetrated the Dawa's network, enabling a very precise reaction by the security services. When the latter moved, it took them only a few days to arrest not only Sadr, but thousands of his supporters, to interrogate them under torture, and then execute nearly everybody. Contrary to Iran, where the Islamists could count on a well-established network of other opposition movements for the organisation and support of popular unrest, the Dawa in Iraq was a poorly organised, loose network of clerics, ordinary members of the Shi'a community, and a few members of the armed forces; it not only proved unable to emulate the revolution of Iran but lacked support from other sections of society. Unsurprisingly, Saddam's security services quickly rendered the entire organisation dysfunctional.[9]

ESCALATION

However, the fuse was now burning: in reaction to Saddam's order for the execution of Sadr, Tehran began broadcasting calls for an overthrow of the government in Baghdad while, always sensitive to any kind of opposition, Saddam Hussein began providing material and financial support to Iranian dissidents. Fermented by age-old enmities between Arabs and Persians, Iranians and Iraqis began attacking each other's officials. By the time the foreign ministers of both countries narrowly survived assassination attempts in April 1980, Iraq had supposedly registered 10 violations of its airspace – including one that went as deep as 45km (28 miles) – and dozens of 'ground attacks on its proper [territory]', and thus concluded that a 'Persian aggression was in the making'.[10]

Who exactly could have violated the Iraqi airspace at that time remains unclear because what was formerly the mighty IIAF, but now officially the Islamic Republic of Iran Air Force (IRIAF), was in tatters and out of condition to do so. Paralysed during the revolution, when it had all of its top commanders arrested and executed, and most of its air bases occupied by groups of armed civilians and non-commissioned officers, it was thrown into complete chaos and on the verge of being disbanded by the new authorities – which even entered negotiations with Great Britain and Turkey regarding a possible sell-out of most of its US-made aircraft, foremost the super-expensive F-14 Tomcats. The condition of the Iranian army was nothing better. Amid an almost complete collapse of order and discipline, and a de facto civil war that raged in Iran between 1979 and 1983, the Iranian armed forces became dysfunctional to a level where they proved unable to operate effectively even against a relatively minor uprising instigated by the Democratic Party of Iranian Kurdistan (KDPI) in north-western Iran in 1979.[11] The condition of the Iranian armed forces only worsened after a failed counter-coup attempt organised by a group of about 400 officers of the air force and the army at Tactical Fighter Base 3 (TFB.3), outside Hamedan, in June 1980; this resulted in the Islamists running another sweeping purge of mid-ranks. By July 1980, the air force that used to have about 100,000 officers and other ranks was down to about 30,000; all the training flights stopped, all the maintenance activity ceased, and the mass of the IRIAF's aircraft were stored – or left to rot wherever parked. Therefore, although it is perfectly possible that some of the border accidents were caused by overzealous officers or local Iranian clergy – often out of eagerness to attract Khomeini's favours – there is little doubt about who instigated the following hostilities.[12]

FERRET-MIGS[13]

From the point of view of Iraq's airmen, the first indication of increasing tensions appeared in April 1980, when the General Command in Baghdad ordered MiG-21Rs of No. 70 Squadron to fly a new series of ELINT-gathering sorties along the border with Iran. Codenamed Operation Ferret, this enterprise was considered so sensitive that only the squadron CO and his deputy were involved. All Ferret sorties were undertaken by single aircraft; usually launched from Rashid AB and flew via Amara to Wahda AB at an altitude of 1,000 metres (3,240 feet), and constant speed of 900km/h.

Although of very poor quality, this rare photograph shows one of the MiG-21Rs of No. 70 Squadron, after an emergency landing (apparently caused by a failure of the hydraulics system). Notably, by this time the jet had received a camouflage pattern similar to that of MiG-21MFs and MiG-21bis in Iraqi service. (Ahmad Sadik Collection)

A flight of Su-20s of No. 1 Squadron, IrAF, deployed to strike a concentration of Kurdish insurgents supported by the authorities in Tehran, at an Iranian army base outside Saradasht, in north-western Iran, on 4 June 1980. (Ahmad Sadik Collection)

The aircraft were unarmed, and always remained 15–20km inside Iraqi airspace, but the pilots were advised to watch out for Iranian interceptors and, in the event of encountering any, dash away at maximum speed while descending to an altitude of 50 metres (164 feet). After refuelling and taking a short break, the pilot would then board his aircraft again, and return via Amara to Ubayda Iban al-Jarrah AB, where the main IrAF Reconnaissance Laboratory had meanwhile been constructed, and this would conduct all the processing and analysis of the results. Processing of the films from the Type-D pods was a particularly laborious and time-consuming issue; early on, the Iraqis would take three to four days to analyse results from just one sortie. Correspondingly, only one Ferret sortie was flown per week. However, as the situation continued to escalate and Iraqi analysts gained more experience, the CO of No. 70 Squadron and his deputy flew two to three times a week. Analysts of the IrAF Intelligence Directorate and the GMID found the results startling: the southern sector of the border was covered not only by Iranian early warning radars, but also dozens of radars supporting air defence systems, such as MIM-23B I-HAWK SAMs, Feuerleitgerät 63 Super Fledermaus radars of Swiss origin used to direct anti-aircraft artillery, and others. The reasons for such activity of the Iranian air defences remains unclear until this very day.

In May 1980, the GMID was informed that the government of Iran had released the Barzani brothers from exile in the USA, and brought them to an army base outside Saradasht, where a large concentration of the Peshmerga was already undergoing training. For Baghdad, the situation was clear: Iranians were preparing a major Kurdish insurgency in Iraq. Correspondingly, the CO of No. 1 Squadron received the order to strike the base in question. The attack was flown on 4 June 1980 by eight Su-20s armed with four FAB-500M-54ShN bombs each. The Iraqis assessed it as highly successful foremost because it caused an uproar from Tehran, which reported six deaths.[14]

Only days later, the IrAF made another foray into Iranian airspace. This time a pair of MiG-21Rs from No. 70 Squadron was sent to fly photo-reconnaissance low along the highway connecting Ahwaz with Abadan, and to overfly the al-Jufair Camp of the Iranian Army – one of the bases of the 92nd Armoured Division. The mission produced photographs showing Chieftain main battle tanks and Scorpion light tanks being moved in the direction of the border.[15]

FROM THREATS TO A WAR

Once again, the background of this movement remains unclear. The only certainty is that the 92nd Armoured Division was largely demobilised at the time, and unable to deploy any of its elements even for limited defensive operations, not to mention an invasion of

A still from a video showing the front section of an Iraqi Mi-25 helicopter gunship. More than 20 were originally ordered and delivered to the IrAF and operated by its newly established Nos. 61 and 66 Squadrons, starting from 1978–1979. However, in the summer of 1980 both units were reassigned to the newly established Iraqi Army Aviation Corps. (Tom Cooper Collection)

A still from an Iranian video (taken from across the border on 7 September 1980) showing the Mi-25 piloted by 1st Lieutenant Athyr Lufty Ahmad Munir, following an emergency landing after it was hit by an Iranian F-14A Tomcat. Reassigned to the IrAAC by that time, Lufty succumbed to his injuries, and became the first Iraqi pilot killed in the coming war with Iran. (Tom Cooper Collection)

Iraq. Indeed, by this time even the GMID was reporting a wholesale collapse and disorganisation of the Iranian armed forces. Together with his growing fears that Khomeini was jeopardising his own political survival, this prompted Saddam to start considering a war against Iran. He neither took that decision easily, nor quickly, and even less so with enthusiasm, but considered it a pre-emptive step exploiting a temporary opportunity to avert the Iranian threat to his rule. The final decision to invade Iran was reached during the meeting of the Ba'ath Party leadership in Abu Ghraib, on 6 July 1980. As a pretext, on 4 September 1980, Baghdad accused Tehran of shelling the vicinity of two villages which Iraq was supposed to receive from Iran in accordance with the Algiers Treaty, and which its army required as a springboard for the planned advance in an eastern direction; when Tehran – preoccupied with the post-revolutionary chaos – ignored demands for their handover, the Iraqi Army, supported by Mil Mi-25 helicopter gunships of the Iraqi Army Aviation Corps (IrAAC), seized both places three days later.[16]

The complete absence of an Iranian reaction then encouraged Saddam to order additional attacks – which in turn started an outright air war. By 7 September 1980, Iraqi commandos – frequently deployed by helicopters – had taken six border posts in the central sector of the border, and then captured a forward COMINT-station

Wreckage of the MiG-21R flown by 1st Lieutenant Jassim Dayekh Muhsin al-Hazza, shot down by IRIAF F-14As on 11 September 1980. Visible in the centre is a burned-out R-13M missile carried for self-defence purposes, and to the left, the fin of the jet. (Farzin Nadimi Collection)

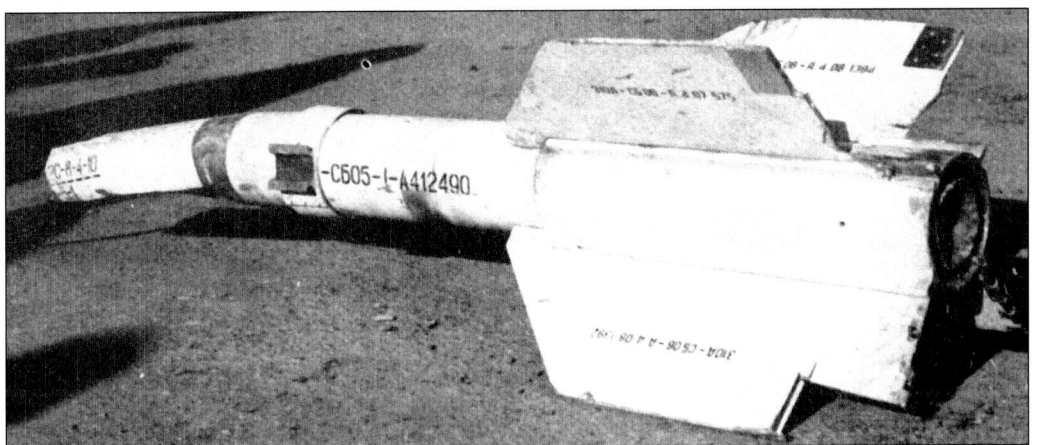

A close-up of the wrecked R-13M missile from Hazza's MiG-21R. In the contemporary press, its wreckage was often erroneously reported as belonging to a 'surface-to-surface missile' – fired either by Iran or Iraq. (Farzin Nadimi Collection)

of the SAVAK, killing dozens of Iranians in the process. When Tehran attempted to order its military to stop Iraqi attacks and incursions, it was forced into the realisation that there was very little left of that military – except for three squadrons of F-4Es at TFB.3, and one each of F-14s at Tactical Fighter Bases 7 and 8 (TFB.7, outside Shiraz, and TFB.8, near Esfahan, respectively). The Deputy Operations of the Iranian air force ordered them into action, but under the condition to attack only targets on Iranian territory. The same afternoon, eight Phantom IIs armed with British-made BL755 cluster bomb units (CBUs) and US-made M117 general purpose bombs bombed the border posts controlled by the Iraqi troops. The F-4s were escorted by a pair of F-14As, one of which detected five Mil Mi-25 helicopter gunships from No. 61 Squadron approaching an Iranian border post in the Zain al-Qaws region. A few minutes later, the lead Tomcat – crewed by Major Kamal Jamshidi, with 2nd Lieutenant Pasha-Pour in the rear – dived to attack and fired about 400 rounds at the rearmost Iraqi helicopter. Iraqi 1st Lieutenant Athyr Lufty Ahmad Munir managed an emergency landing on the Iraqi side of the border before succumbing to his injuries. Iraq had thus lost the first air combat of the coming war.[17]

A day later, the IRIAF lifted the embargo on missions inside Iraq and TFB.3 received the order to bomb a bridge outside Sarpol-e Zahab. This time, one of the involved F-4Es was shot down by an Iraqi 9K32M Strela-2M (ASCC/NATO-codename 'SA-7 Grail') man-portable air defence (MANPAD). The crew ejected safely and was picked up by an IRIAF Bell 214C SAR-helicopter two days later, but this was then shot down by friendly fire while passing Iranian positions near Qasr-e Shirin, killing everybody on board. Starting from 9 September 1980, F-4Es from TFB.3 and F-14As from Tactical Fighter Base 8 (TFB.8), near Esfahan, began flying CAPs over the combat zone, attempting to intercept some of the Iraqi helicopters involved in deploying special forces and delivering air strikes on the remaining Iranian positions. During the day, Iranian Phantom crews claimed to have shot down or damaged four Mi-8s, all by gunfire. This in turn prompted the IrAF to the scene. By that point in time, the Iraqi jets were flying only reconnaissance sorties inside Iranian airspace, but now the High Command in Baghdad ordered them to conduct air strikes on enemy ground forces and to fly CAPs above their own troops. A clash between Iranian and Iraqi fighters thus became inevitable and, according to Iranian sources, IRIAF F-14As then shot down an Iraqi 'Su-20M' over the Sarpol-e Zahab area, using an AIM-54A Phoenix missile, late in the afternoon of 9 September 1980. While the Iranians confirmed this 'kill' by intercepting enemy radio communications, the Iraqis insisted that they suffered no such losses. It was only in 1990 that the IrAF did admit a 'similar' loss, even though they dated it on 14 September, when the Su-22 piloted by the CO No. 44 Squadron, Major Noubar Abdel-Hamid al-Hamadani 'blew up' while underway on a mission of visual reconnaissance, without any kind of warning. Hamadani was declared 'missing in action' and never heard of again; because by that time the mass of helicopter units involved had been reassigned to the IrAAC (see below for details), losses of their machines were not counted in any of the related IrAF reports, and Hamadani thus became the first IrAF pilot lost in the war with Iran. Tragically, he was not the last. Two days later, on 11 September 1980, No. 70 Squadron – the jets of which had meanwhile flown dozens of reconnaissance missions inside Iranian airspace without ever being caught by the enemy – ran out of luck and lost one of its precious MiG-21Rs. The jet piloted by 1st Lieutenant Jassim Dayekh Muhsin al-Hazza was shot down while underway 'north of Abadan' and killed. The IrAF Intelligence Directorate assessed this loss as caused by 'Iranian SAMs': however, none were actually nearby, and the MiG was felled by another F-14A, using an AIM-54A Phoenix long-range air-to-air missile. Just as in the case of Hamdani's Su-22, two days earlier, the Iraqi electronic warfare units failed to record emissions from Iranian AWG-9 radars.[18]

Constrained by the short endurance and short-ranged weaponry of their jets, Iraqi fliers were less successful in intercepting IRIAF fighters. Certainly enough, 1st Lieutenant Sadiq, a MiG-21MF pilot from No. 11 Squadron, fired two R-13Ms at an F-4E which was underway about three kilometres inside Iraqi airspace and attacking elements of the 6th Armoured Division outside Khanaqin on 8 September 1980. The Intelligence Department IrAF credited him with a confirmed kill after finding out that the Iranian pilot, Major Mahmoud Eskandari, managed to nurse the badly damaged jet back over the border before ejecting safely (his back seater, 1st Lieutenant Ali Ilkhani, was killed). However, the Iranians credited this loss to a combination of their own heavy machine guns and ZU-23 anti-aircraft guns. The IRIAF did lose two Northrop F-5E Tiger IIs from Tactical Fighter Base 4 (TFB.4/Vahdati, outside Defzul), in the following days, but both of these were to Iraqi Army ground defences.[19]

Less lucky were a pair of MiG-23MS pilots on 14 September 1980. They intercepted a Bell 214 helicopter carrying the Iranian President Abolhassan Banisadr and Colonel Javad Fakoori (who at the same time acted as the Iranian Minister of Defence, the Chief of the Joint Chiefs of Staff of the Iranian armed forces, and the Chief of Staff IRIAF), as it was touring the border in the Ilam area, escorted by two Bell AH-1J Cobra attack helicopters of the Islamic Republic of Iran Army Aviation (IRIAA). The Iranian pilots dropped down to low altitude and flew hard evasive manoeuvres, causing at least two Iraqi missiles to miss before the MiGs distanced, probably out of fuel.

As the Iraqi Army continued gradually escalating its pressure along the common border, resulting in additional clashes – and another, this time unsubstantiated IrAF claim for downing an IRIAF F-4E – and after securing financial and political support from Kuwait and Saudi Arabia, on 17 September 1980 Iraqi President Saddam Hussein announced the cancellation of the Algiers Treaty with Iran, and claimed sovereignty over the Shatt al-Arab waterway. The way into a war with Iran was now open.[20]

7

IRAF OF 1980

By the summer of 1980, the Iraqi Air Force had grown to about 38,000 officers and other ranks, organised as described in Table 4. Its members considered themselves the elite of the Iraqi armed forces, hardened by combat experience against Israel, Barzani's Kurds and the Iranians. Nevertheless, a superficial glance at its doctrine, strategy, organisation and equipment as of August and September that year indicated numerous weaknesses. Regardless of how many new aircraft had been ordered in previous years, the mass of them were still not in service; indeed, not even in Iraq. The backbone of the fighter fleet was still the MiG-21, and thus the overall fighting capability of the entire force was only marginally better than in 1973. Unsurprisingly, even the IrAF strategic plan – whether for a war with Israel or with Iran – was defensive by nature; the air force was trained and equipped to expect, accept, and survive the first strike (regardless of whether it was on political, military or industrial targets) and only then to hit back. However, all the investment in recruitment and training of additional personnel – and thus the massive expansion of the force in the period 1976–1980 – was not in vain. On the contrary, despite a never-ending flow of reports about the IrAF being critically short of qualified personnel, over the following years it was to show itself to be perfectly capable of rapidly acquiring and introducing to service a large number of entirely different combat aircraft and other equipment. This would have been impossible without the establishment and massive expansion of the Directorate of Air Training during Sha'ban's tenure – and in the years before. Moreover, while Western observers continued describing the IrAF as 'Soviet trained' for decades later, the force was organised and operated following a doctrine that was a mix of its own experience and Western theories. Moreover, and at least in terms of its uniforms and traditions – and despite most of its aircraft being 'made in the USSR' – the Iraqi Air Force continued drawing on strong relations to the Royal Air Force; as described above, the cooperation with the Soviets was always limited to arms purchases. As well as help with conversion training and the introduction to service of new aircraft types – and although the IrAF continued sending large numbers of its officers and other ranks to the Soviet Union and Czechoslovakia for training – it also continued sending its personnel for training in Great Britain, France, India, and Pakistan, while large groups of pilots from the Indian Air Force (IAF) served at the Air Force Academy at Sahra AB. With the IAF heavily leaning on strong RAF-related traditions, these remained strong within the IrAF too, as was obvious from its organisation, uniforms, and ranks, but especially in terms of doctrine and tactics. Falah Hassan, who joined the Academy in the mid-1970s, drew an interesting comparison:

> Our training was conducted in a mix of languages, including English, Russian, and even some Hindi. Indian instructors taught us to fly Czechoslovak-made L-29s and L-39s in our advanced courses. The Russians then trained us in regards of safety, following basic rules and keeping people alive. However, the Indian training was much more aggressive: their indoctrination made us confident in our abilities, encouraged us to be creative and courageous.[1]

In other words, regardless of their being no direct British presence and irrespective of its Soviet equipment, the IrAF considered itself a 'Western-style' air force, and its personnel was confident of being well on the way to not only solving all the problems it was facing, but, indeed, on the best way to a brilliant future.

AIR DEFENCE

As of August and September 1980, the Kari IADS was still non-operational and available radar systems were unsuitable for detecting low-flying aircraft, especially against the backdrop of the Zagros mountain range stretching on the north-west–south-

A MiG-21MF of the IrAF seen on landing. Notable is the typical 'CAP-configuration' of the late 1970s and early 1980s, including three 400-litre drop tanks (one on the centreline, and one each of outboard underwing pylons), and two R-13M missiles (inboard underwing pylon). (Farzin Nadimi Collection, via Tom Cooper)

The second batch of MiG-21bis delivered to Iraq was assigned to No. 37 Squadron. They were recognisable by a rather 'irregular' version of their standardised camouflage pattern. This example (serial deleted by the Iraqi military censor) nicely shows one of its contemporary weapons configurations: drop tank under the centreline, R-13M missile on the inboard underwing station and the old R-3S missile on the outboard underwing stations. The latter was withdrawn from service in 1979, but still used for training purposes. (Tom Cooper Collection)

east axis on the Iranian side of the mutual border. French-made Tigre-G low-altitude radars were on order; while promising to be capable of remaining operational in the face of severe electronic countermeasures, they were yet to be delivered. Therefore, and because of the speeds of modern fighter-bombers, in the event of war the Directorate of Air Defence was forced to operate in an old-fashioned style, by keeping a large number of interceptors airborne and flying combat air patrols (CAPs) to cover all the possible enemy ingress routes, and to do so from dawn until dusk.[2]

The combination of intensive operations and a minimal maintenance requirement was the strength of the MiG-21 family, and this was precisely the reason why the type still formed the backbone of the IrAF at that time: over 100 aircraft of three older variants – MiG-21MF/R/UM – were operated by a total of five units, with a significant stock of aged MiG-21F-13s, MiG-21FLs and MiG-21PFMs in storage and kept back as reserve. Unsurprisingly, during the spring and summer of 1980 this fleet was further expanded as, once preparations for a war with Iran were initiated, the High Command IrAF ordered No. 17 OCU (formerly No. 17 Squadron) and 27 OCU to convert into operational units, even though equipped with older MiG-21FLs and MiG-21PFMs. Meanwhile, in late 1979, the first batch of 18 MiG-21bis was delivered from the USSR; all turned out to have been former MiG-21SMTs of the Soviet air force, overhauled and upgraded to the MiG-21bis standard, but the Iraqis were now in such a rush that they were accepted and pressed into service with the newly established No. 47 Squadron. By August 1980, another 36 MiG-21bis were in the country, enabling the re-equipment of the veteran No. 14 Squadron, and the establishment of another new unit, No. 37 Squadron (the latter was still working up as of September 1980). At least as important as the acquisition of the MiG-21bis was the fact that the Soviets delivered them together with 330 additional R-13Ms and 240 brand-new R-60MK infra-red homing short-range air-to-air missiles (ASCC/NATO-codename 'AA-8 Aphid'). This small, light and nimble weapon was to prove highly reliable and popular in service and was an excellent complement to the R-13M. Finally, in August 1980, France delivered the first batch of Matra R.550 Magic infra-red homing short-range air-to-air missiles – and the technicians of No. 47 Squadron, the unit responsible for air defence of the construction site of Saddam AB, promptly scrambled to modify several of their MiG-21bis to carry this weapon.

Deliveries of R-13Ms, R-60s and R.550s enabled the IrAF to finally withdraw from service all of the remaining R-3Ss – with one exception; this weapon was retained by the sole interceptor unit flying types other than the MiG-21; the MiG-23MS-equipped No. 39 Squadron. The reason was that the flight-testing of newer R-13Ms revealed that its motor developed so much smoke that when fired from one of two under-fuselage stations it caused the engine to surge. Therefore, while all the MiG-23MS had all four of their weapons hardpoints rewired for deployment of the R-13Ms, in service these were installed only on underwing pylons. Having no other options to hand, the Iraqis were forced to continue installing the near-useless R-3S on under-fuselage hardpoints of their MiG-23MS – all provided the type was deployed for intercept purposes; facing shortages of fighter-bombers, while preparing the invasion of Iran, the High Command IrAF decided to relegate No. 39 Squadron to ground attack tasks.[3]

In regards of tactics, the IrAF interceptor force of mid-1980 was on a par with most Western air forces. Pilots depended on ground control to provide an initial intercept vector, but – except under very specific circumstances – were largely free in developing their subsequent operations, and they were trained to fly and fight in open, spaced formations, elements of which provided mutual protection. Their primary limitations were all imposed by their equipment. Both the MiG-21MF/bis and MiG-23MS had only rudimentary radars, very poor radar warning receivers (RWRs), and offered a very poor view outside their cockpits, and MiG-21s were always short on fuel. The issue of RWRs was of especially crucial importance because it was only years later that the Iraqis realised that the systems in service during the early 1980s regularly failed to warn them about the presence of F-14s. Furthermore, the Iraqis were absolutely unaware of the fact that the Iranian air force was equipped with US-made enemy-IFF interrogators, capable of detecting enemy aircraft without actually activating radars. Finally, the only aspect in which Soviet influence was felt within the IrAF was the emphasis on deployment of air-to-air missiles in combat; like the contemporary Soviet tactics, so also the Iraqi interceptor-tactics emphasised deployment of missiles, which required less-precise aiming, while shortening the duration of every engagement, in turn making their own aircraft less susceptible to coming under enemy counterattack. Both MiG-21s and MiG-23s were equipped with internal GSh-23-2 twin-barrel 23mm automatic cannons and gunnery training was conducted regularly, but pilots rarely employed them in combat.

SUSPENSION OF PROJECT 84[4]

At least in theory, during the summer of 1980, the IrAF was due to introduce to service two new interceptor types. Unexpectedly, this process was rudely interrupted, because of malicious Soviet business practices. As mentioned above, through 1979 the IrAF placed orders for a total of 44 MiG-25s; codenamed Project 84 (because '84' was the air force's designation for the MiG-25), the service introduction of this type required comprehensive preparations, including not only the training of necessary personnel, but also the expansion and hardening of several air bases to make them suitable for operation of this large and heavy new type, and the construction of additional fuel depots suitable for the type of kerosene used by the Mach 2.83-capable jet. The Soviets delivered the first batch of eight MiG-25PU two-seat conversion trainers and MiG-25RB reconnaissance bombers by transport aircraft to Rashid AB in late January 1980. From there, they were trucked to Taqaddum for assembly and test-flights before being officially accepted by the IrAF. By this time, the first group of 16 of Iraqi pilots (all veteran MiG-21MF/R operators) had completed a conversion course at the 802nd Training Air Regiment at Krasnodar AB and the 615th Training Air Regiment of the 148th Training Centre, at Sevasleika AB, in the USSR and were back in the country. Together with a large number of technicians (Team 841 was responsible for first line maintenance, Team 842 for the second line), they formed the core of No. 87 Squadron – commanded by Lieutenant Colonel Abdullah Faraj al-Azzawi – which initially acted as an OCU for the type.[5]

The first Iraqi MiG-25 unit was still working up to the time the first five MiG-23MFs were delivered by Soviet transport aircraft to Rashid AB. The first two groups of pilots and ground personnel for them underwent conversion training in the USSR in early 1980, and were eager to accept their new jets, which, apparently, were to finally fulfil Soviet promises made for the MiG-23MS. Their S-23 weapons system was centred on the Sapfir-23D-III pulse radar with a maximum detection range for fighter-sized targets underway at medium or high altitudes of about 45km (24nm), or 10–20km (5.3–10.6nm) for low-flying targets. Of high importance were their primary weapons: R-23R semi-active homing, medium-range air-to-air missiles (ASCC/NATO-codename 'AA-7A Apex' and 'AA-7B Apex', respectively). These could be deployed against the front hemisphere of the target and against targets underway at an altitude down to about 1,000 metres (3,280 feet), over a maximum effective engagement range of between 15km (8nm) for the R-23T and 25km (13.5nm) for the R-23R. Like MiG-25PUs and RBs before them, the first five MiG-23MFs were brought to Taqaddum for assembly and test-flights.[6]

However, what the Iraqis discovered while inspecting the aircraft before officially accepting them caused a shock: while Baghdad had ordered, signed a contract for, and paid for newly built aircraft, and the Soviets agreed to deliver exactly that, these five MiG-23MFs turned out to have been only partially-overhauled secondhand jets. The variant was out of production and the Soviets could not deliver any newly manufactured examples. Moreover, Moscow had decided not to deliver any R-23Ts to Iraq. Unsurprisingly, the IrAF flatly refused to accept them; all were trucked to Wallid AB and stored pending further negotiations with Moscow. Wary because of this experience, officers of the Directorate of Logistics were particularly cautious during the next Soviet delivery, that of the first MiG-25 interceptors. As in the case of the MiG-23MFs, the Iraqis ordered and paid for newly built MiG-25PDs, including such improvements as the RP-25 Smerch A2 radar (with a maximum detection range of about 70km/37nm for fighter-sized targets), the TP-26Sh infra-red search and track (IRST) system, the capability of carrying a 5,000-litre drop tank under the centreline, and compatibility with R-60MK short-range air-to-air missiles – in addition to their standard armament consisting of R-40RD semi-active radar homing (SARH) and R-40TD infra-red homing medium-range missiles (ASCC/NATO-codename 'AA-6C Acrid' and 'AA-6D Acrid', respectively). Instead, the Soviets delivered second-hand MiG-25Ps with RP-25 Smerch A1 radars, no IRST, unable to carry drop tanks, and lacking compatibility with R-60s. The IrAF found itself without an option but to refuse to accept these jets too. Whether any were assembled and test-flown before the deal was stopped remains unknown; the only certainty is that all ended up being stored at Wallid AB, pending a solution to the resulting dispute. The service introduction of MiG-23MFs and MiG-25PDs thus experienced an unplanned suspension; before long, it was to experience another, much more massive delay.

STRIKE ASSETS

As of August and September 1980, the spear tip of the IrAF's strike capacity was the Strategic Brigade IrAF, which still included No. 10 Squadron operating six old Tu-16 bombers, and Nos. 18 and 36 Squadrons operating a total of 15 supersonic Tu-22 bombers. All were now old and showing their age, but they represented the only weapons in the Iraqi arsenal capable of reaching places in Iran as distant as Tehran, Qom or Esfahan. On the basis of experience from the Second Iraqi-Kurdish War, all had now been upgraded through the installation of Sirena-2 RWRs, Yolka swept noise jammers (ASCC/NATO-codename 'Chip Thorn'), and ASP-2B chaff and flare dispensers.[7]

As of mid-1980, the new 'star' of the IrAF fighter-bomber fleet was planned to become the brand-new version of the Su-20/22 family: the Su-22M. Of course, in the light of their recent experiences with MiG-23MFs and MiG-25s, the Iraqis were extremely cautious about accepting the first 18 jets of the new variant into service, and the Soviets took care to accompany them with a sizeable team of specialists. However, the new Sukhois arrived equipped exactly as specified in the related order. They had a terrain-following radar installed low in the front fuselage, and compatibility with a wide range of weapons, including – in addition to usual general purpose and incendiary cluster bombs, unguided rockets and R-13M air-to-air missiles for self-defence – the latest Kh-25 and Kh-29 guided air-to-ground missiles (ASCC/NATO-codenames 'AS-10 Karen' and 'AS-14 Kedge', respectively). Moreover, all Su-22Ms delivered to Iraq could be equipped with the SPS-141 electronic warfare pod; only a delay in the development of this system resulted in its delivery being postponed to 1981. The new Sukhoi variant first entered service with Hurrya-based No. 5 Squadron, replacing Su-7BMKs, which – supported by a group of Soviet specialists – quickly worked up to initial operational capability.[8]

Of the older Sukhois, Su-22s were now operated by Nos. 44 and 109 Squadrons. However, instead of being kept together, one of these was home-based in northern Iraq and the other in southern Iraq; pilots of No. 44 Squadron were trained for CAS and COIN operations against Barzani's Peshmerga, while those of No. 109 Squadron became the first Iraqi fliers to receive dedicated training in anti-ship attacks. In turn, this excluded their cooperation or mutual support. Unknown to the Iraqis, the condition of No. 1 Squadron's Su-20s was soon to become much poorer. In July 1980, one of its jets crashed shortly after take-off, killing Captain Jalal Rashad Arsalan. The investigation quickly revealed a major engine problem, prompting the ground crews to inspect all the other aircraft – and found out that these were about to suffer from the same issue. Eventually, the IrAF was left with no choice but to request Moscow

Staff of No. 10 Squadron, equipped with Tu-16 bombers, seen in front of their administration building at Taqaddum AB in the summer of 1980. Notable is the crest of this unit applied on the wall to the rear. (via Ali Tobchi)

A Su-22M of No. 5 Squadron (apparently wearing the serial number 2026), seen at Hurrya AB in the summer of 1980. Notable is the camouflage pattern consisting of dark brown and dark green. (Tom Cooper Collection)

Alwan al-Abossi in the cockpit of a Su-20 of No. 1 Squadron, IrAF, in August 1980. Following the death of one of the pilots from this unit, the entire fleet was grounded, pending an investigation by a team of Soviet experts. (Alwan al-Abossi Collection)

Captain Jalal Rashad Arsalan, killed while piloting a Su-20 of No. 1 Squadron in July 1980. The investigation into this accident led to the grounding of the Su-20 fleet, severely reducing the striking power of the entire air force at a crucial moment in time. (via Ali Tobchi)

to deploy a team of specialists to find a solution. With the Soviets being slow to react, and then with their work, all the remaining 16 aircraft of the unit were grounded on 4 September 1980. No. 1 Squadron flew not a single sortie for the next 20 days and remained of only limited use for months afterwards.[9]

This 'loss' of a precious fighter-bomber squadron could not have come at a worse moment in time – and it put an additional burden upon other units of this kind, of which there were few. The situation became critical enough for the High Command IrAF to order the Su-7BMK equipped No. 8 OCU to be prepared for combat operations when the time for the invasion of Iran came. It remains unclear if this was also the case with No. 20 OCU, a little-known unit reportedly equipped with Su-22UMs and active at least during 1979–1981.

As much as many of the IrAF commanders preferred Soviet-made over Western-made aircraft, they also preferred Sukhois to MiGs. The IrAF MiG-23BN fleet was thus often deemed an 'ignored' asset; indeed, rumours were making the rounds – much to the offence of officers serving in the units operating this type – that only those suspected of having connections to the Dawa Party were assigned to them. As subsequent developments were to show, Iraqi MiG-23BN pilots were not only patriotic, but also extremely courageous. Home-based at Abu Ubayda Ibn al-Jarrah AB (near al-Qut) and Ali Ibn Abu Talib AB, respectively, Nos. 29 and 49 Squadrons were relatively close together and capable of providing mutual support. Even if having a rather austere avionics outfit, their jets proved capable of carrying the same war-loads as Su-20/22s over a slightly longer range – and, after experiencing so many problems at earlier times, nearly all of about 30 surviving jets were fully operational.

TRANSPORT FLEET

Perhaps the most important, yet most ignored, of the Iraqi orders for Soviet aircraft from 1978 was that for Il-76MDs. The reason was related to the combination of payload, speed, and high reliability of this aircraft: it could lift a maximum payload of 40,000kg (88,000lbs) to an altitude between 9,000m and 12,000m (29,530–39,370ft) and then cruise at a speed of 750–800km/h (405–432 knots) over a range of 5,000km (3,107 miles). Moreover, the Il-76MD was equipped with front and rear cargo doors, had a cargo-roller floor, two 3,000kg (6,500lbs) winches, and two roof cranes with a total capacity of 10,000kg (22,000lbs) – all of which made their loading and unloading much easier, and enabled the transport of oversized vehicles or equipment. Indicating the total capability of Il-76MDs, it is sufficient to say that the type could haul all the cargo and

A brand-new Il-76MD of the IrAF, sporting the colours and insignia of Iraqi Airways, seen in the early 1980s. (Martin Hornliman/Milpix Collection)

personnel the old IrAF transports had flown to Egypt and Syria of 1973 in just 28 flights, and all within one day.[10]

Initially 'limited' to 18 aircraft, the Iraqi acquisition resulted in the delivery of six Il-76MDs through 1978–1979. Additional examples followed during the 1980s, and by 1990 the fleet was increased to 40–42 aircraft. Although all wore the livery of Iraqi Airways, the first six Il-76s were operated exclusively by the air force; the first group of flight-crews and ground personnel – all with previous experience from operating An-12s and An-26s – completed their conversion courses with the 610th Training Regiment at Ivanovo AB in the USSR in late 1978. They established No. 33 Squadron. About a year later, the second unit equipped with the type came into being: No. 43 Squadron. The service entry of Il-76s rendered the miscellany of old – and worn out – transports surplus, and they were all retired. Only the old No. 23 Squadron continued flying a mix of about two dozen An-12 and An-24 medium transports (the former of which saw heavy utilisation well into the 1980s), while an unknown unit flew Antonov An-2 biplane transports – which were in service with the Air Force Academy and in Iraqi agricultural aviation.[11]

HELICOPTER FLEET

The IrAF helicopter fleet grew immensely during the 1970s, until it counted more than 200 Mil Mi-4s, Mi-6s, Mi-8s, Mi-25s, Alouette IIIs, Gazelles and Super Frelons. In 1980, the General Command in Baghdad thus decided to redistribute this force in order to enable its better specialisation and closer cooperation with the other branches of the armed forces. Correspondingly, the mass

Aerial photograph of the old K-2 airfield from November 2003. Used by the IrAF as a forward operating base since the 1930s, it was handed over to the IrAAC in the summer of 1980. (Photo by Martin Rosenkranz)

Pilots of No. 101 Squadron with Brigadier General Farouk Faraj in 1977, shortly after the unit was established; by 1980, the sole Iraqi asset operating Super Frelon helicopters was subordinated to the Iraqi Navy. (via Ali Tobchi)

Table 4: IrAF ORBAT, September 1980				
Chief of Staff: Major General Mohammad Jassam Hannash al-Jaboury (1978–1984)				
Deputy Operations: Brigadier General Muhammad Saliman Hamad (1978–1984) Deputy Training: ? Deputy Air Defence: Brigadier General Ahmad Khayri Deputy Logistics and Planning: Brigadier General Osama al-Yawer Director Intelligence Directorate: Lieutenant Colonel Fawaz as-Selo				
Unit	Base	Equipment	Remarks	Commanders
Direct-Reporting Units				
No. 3 VIP Transport Squadron	Muthenna IAP	JetStar, Falcon 20	est. 1935	
No. 4 VIP Transport Squadron	Muthenna IAP	SE.316B, Mi-8,	est. 1936; re-est. 1950 as helicopter unit equipped with Wessex; re-equipped with SE.316B in mid-1979	
Flight Leaders School	Rashid AB	Hunter F.Mk.59A/B, Jet Provost	est. 1966; fighter-weapons school	Col. Muhammad Nabil Ahmed Ayoub
No. 33 Squadron	Baghdad IAP	Il-76MD	est. 1978	
No. 43 Squadron	Baghdad IAP	Il-76MD	est. 1979	
1st ADS (Centre), HQ at-Taji				
No. 11 Squadron	Rashid AB	MiG-21MF	est. 1961; re-equipped with MiG-21MF in 1972	
No. 23 Squadron	Rashid AB	An-12, An-24	est. 1965, partially re-equipped with An-24 in mid-1970s	
No. 70 Squadron	Rashid AB	4x MiG-21R, 12x MiG-21MF	est. 1973	
SAR Flight	Rashid AB	2x Mi-8		
No. 145 Missile Brigade	Baghdad IAP	6x SA-2, 6x SA-3		
No. 10 Squadron	Taqaddum AB	9 Tu-16	est. 1960	Lt. Col. Adel Abdul Hamid Uthman
No. 18 Squadron	Taqaddum AB	6 Tu-22	est. 1973; incl. aircraft 1110, 1111, 1112, 1113 (non-operational) 1114, 1115 and 1116	Maj. Ibrahim Muhammad Ali al-Farkahi (KIA October 1980)
No. 36 Squadron	Taqaddum AB	7 Tu-22	est. 1975; incl. aircraft 1117, 1118, 1119, 1120, 1121 and 1122	Lt. Col. Luay at-Tabaqjaly
No. 39 Squadron	Taqaddum AB	20x MiG-23MS	est. 1974; detachment at Wallid AB	Lt. Col. Muhammad Ali Dawood al-Jaboury
No. 146 Missile Brigade	Taqaddum AB	2x SA-2, 4x SA-3		
SAR Flight	Taqaddum AB	2x Mi-8		
No. 67 Squadron	Bakr AB	MiG-23MF	still working up as of August 1980, fully operational only in 1982	Maj. Daham Shia'a Alwan, (KIA October 1984)
2nd ADS (West), HQ Wallid AB				
			under construction as of August 1980, no permanently assigned units	

Table 4: IrAF ORBAT, September 1980				
3rd ADS (South), HQ Nasiriyah				
No. 14 Squadron	Ali Ibn Abu Talib AB (Nasiriyah)	MiG-21bis	est. 1966 with MiG-21FLs; re-equipped with MiG-21PFM in 1969; re-equipped with MiG-21bis in 1979–1980	Maj. Qaysar Ibrahim Ramzi (KIA November 1981)
No. 49 Squadron	Ali Ibn Abu Talib AB	MiG-23BN		Lt. Col. Fahd Abdul Baqi al-Uqaily (1979–1983)
SAR Flight	Ali Ibn Abu Talib AB	2x Mi-8		
No. 29 Squadron	Abu Ubayda Ibn al-Jarrah AB (al-Qut)	MiG-23BN	est. 1966; re-equipped with MiG-23BNs in 1975–1976	Maj. Rashid Hamad as-Sa'aydon (KIA September 1980); Maj. Jabbar Hammad ad-Dulaimy (KIA September 1980); Maj. Muhammad Mudher al-Farhan
SAR Flight	Abu Ubayda Ibn al-Jarrah AB	2x Mi-8		
No. 109 Squadron	Wahda AB (Basra)	Su-22M	est. 1978	Maj. Hisham Ismail al Barbouti (1979–1981)
No. 148 Missile Brigade	Wahda AB	3x SA-2, 5x SA-3		
SAR Flight	Wahda AB	2x Mi-8		
No. 149 Missile Brigade	Um Qassr Naval Base	3x SA-3		
4th ADS (North), HQ Kirkuk				
No. 1 Squadron	Hurrya AB	15x Su-20	est. 1931; re-equipped with Su-20 in 1973–1974; all aircraft non-operational as of Sep 1980	Maj. Khaldun Khattab al-Bakr
No. 5 Squadron	Hurrya AB	18x Su-22M	est. 1953; re-equipped with Su-7BMKs in Feb 1969; re-equipped with Su-22M in 1978	Maj. Mohammed Hamid Taha (POW Sep 1980)
No. 37 Squadron	Hurrya AB	MiG-21bis	est. 1979; working up as of Sep 1980	
No. 44 Squadron	Hurrya AB	Su-22	est. 1977	Maj. Nubar Abd al-Hamid Hamadani (KIA Sep 1980)
No. 47 Squadron	Hurrya AB	MiG-21bis	est. 1979; operational as of Sep80; redeployed to al-Wallid AB in 1990	Maj. Mohammad Sakran (1979–1981)
No. 195 Missile Brigade	Hurrya AB	3x SA-2E, 3x SA-3		
SAR Flight	Hurrya AB	2x Mi-8		
No. 9 Squadron	Firnas AB	MiG-21MF	est. 1959; detachment at Abu Ubaida AB	Maj. Kamal Abdal Sattar al-Barzanji (1980–1981)
SAR Flight	Firnas AB	2x Mi-8		
Directorate of Air Training				
Air Force Academy	Sahra AB (Tikrit)	Zlin, L-29, L-39, MiG-15UTI	est. 1970–1973	Col. Tahir Saleh Tawfiq at-Tikriti
No. 6 OCU			est. 1954; status unclear as of 1980	
No. 7 OCU	Abu Ubayda Ibn al-Jarrah AB, then Baghdad IAP	Su-7BMK	fighter-weapons school in 1970s; re-equipped with MiG-21PFM and activated as combat unit in August 1980, with few MiG-21F-13s in reserve	

Table 4: IrAF ORBAT, September 1980

No. 8 OCU	Abu Ubayda Ibn al-Jarrah AB	Su-7BMK, Su-7UM?	est. 1958 as Il-28-squadron; re-equipped with Su-7s from No.1 Squadron, 1975 and acting OCU for Su-22 squadrons	Maj. Ahmed Nayef al-Juboury
No. 17 OCU	Sahra AB	MiG-21FL/UM	est. 1966; MiG-21-OCU until August 1980 then operational squadron for air defence of Sahra AB; redeployed to Abu Ubaida AB in November 1980	Maj. Muwafaq ad-Dayny
No. 20 OCU		Su-22UM	reported as OCU in 1978–1979; status unknown	
No. 21 OCU	Sahra AB	SA.342	est. 1980; helicopter pilot training	
No. 27 OCU	Abu Ubaida Ibn al-Jarrah AB, Tikrit AB, al-Wallid AB, Sa'ad AB	12x MiG-21FL/PFM, 12x MiG-21UM	est. July 1980 as MiG-21 OCU at Rashid AB; redeployed to Sahra AB in Sepember 1980	
No. 59 OCU	Taqaddum AB	MiG-23UB	est. 1974	
No. 79 OCU	Saddam AB	Mirage F.1EQ	est. 1980 but still working up; personnel in France; reorganised as No. 79 Squadron in 1981	
No. 87 OCU	Tammuz AB	MiG-25PU	est. 1980–1981; working up, personnel in USSR; reorganised as No. 87 Squadron in 1981	
SAR Flight	Sahra AB	2x Mi-8		

of helicopters with assault and attack roles were reassigned to the newly established Iraqi Army Aviation Corps and reorganised into three wings of five to six squadrons each. The IrAAC took over not only the mass of IrAF helicopters, but also several bases originally constructed for the air force. These were at-Taji, north of Baghdad, and Iskandariya, south of the Iraqi capital, and two old forward operating bases constructed along the K-Pipeline; K-1 and K-3. In similar fashion, No. 101 Squadron, which operated Super Frelons armed with AM.39 Exocet anti-ship missiles, was reassigned to the Iraqi Navy. Pending construction of a small air base outside the main base of the Iraqi Navy at Umm al-Qasr, the unit was home-based at Ali Ibn Abu Talib AB. This left the IrAF with only a handful of helicopter squadrons, mostly operated for communication, light transport, utility or search and rescue purposes.[12]

Short on bombers and fighter-bombers, the IrAF was forced to press into service not only old MiG-21FLs and MiG-21PFMs but also Su-7BMKs and Hunters, and to convert numerous training units into operational squadrons during preparations for the invasion of Iran in the summer of 1980. Surviving Hawker Hunters – veterans of several wars with Israel and the Kurds, now operated by the Flight Leaders School, IrAF – were thus to perform their 'swan song' during another big war. (Albert Grandolini Collection)

WINGS OF IRAQ VOLUME 2: THE IRAQI AIR FORCE 1970–1980

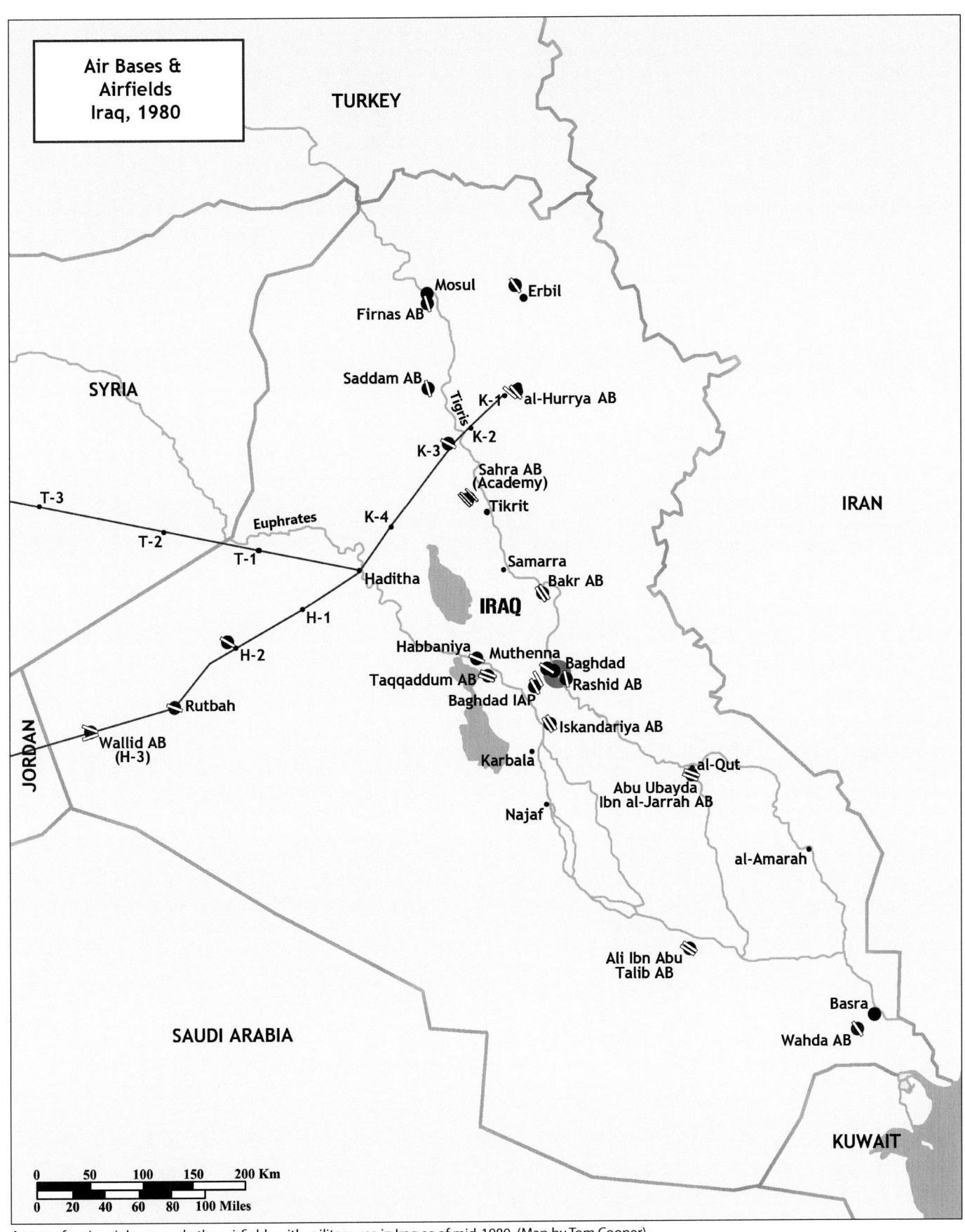

A map of major air bases and other airfields with military use in Iraq as of mid-1980. (Map by Tom Cooper)

8
ECHO OF QADESSIYA

Once Saddam Hussein and the RCC made the decision to invade Iran, it was up to the General Command of the Iraqi Armed Forces in Baghdad to convert the idea into military action. The resulting overall plan was to open this operation with an air strike on selected air bases in western and central Iran. The Iraqi Air Force thus played a crucial role – at least in the first step of the Iraqi invasion.

IRAF PLANNING

The GMID assessed the IRIAF as being theoretically superior to the IrAF, and on alert, but in a very poor condition. Iraqi military intelligence knew that Iranian air bases were all heavily fortified and thus impossible to completely destroy with available aircraft. Therefore, the IrAF received the order to fly only two waves of air strikes and to aim for runways and fuel depots. The logic was that this would be enough to hamper an IRIAF reaction for the first 48 hours of the war, and thus equalise its advantages by a margin. On the ground, the Iraqi Army was only to guard the northern sector of the border or make shallow penetrations into Iran in the central sector. Its emphasis was in the south, where mechanised formations were to make a quick dash of between 20km and 75km (12.5 and 46.5 miles, respectively) deep into Khuzestan Province in south-western Iran, up to the narrow passes along the mountain ranges stretching from north-west to south-east. This was in accordance with the official explanation for the invasion, in which Iraq had tried to solve the dispute with Iran through diplomacy and international law, but Iran ignored these attempts, violated Iraqi airspace 249 times over the previous 12 months, blocked the Shatt al-Arab waterway, bombarded Iraqi cities and villages with artillery, and massed troops along the mutual border.[1]

JABOURY'S DISAGREEMENT

Interestingly, at least from the point of view of the leading Iraqi airmen, this plan was doomed to fail. The most massive mistake of both Saddam and the General Command in Baghdad was to order their fliers into a strategic defensive; after the opening air strike, the IrAF was to pull back and wait for an Iranian counterattack, and then seek to impose attrition over its home turf.[2]

This idea left Jaboury and his deputies deeply concerned, and actually in opposition at least to the General Command, if not even to Saddam. Their reasoning was purely military by nature. Although intelligence reported no activities by IRIAF F-14s during the border skirmishes through September 1980, Operation Ferret had already revealed a clear picture of the Iranian air defences. This made both the GMID and the High Command IrAF unhappy; they still felt that their 'picture' of the Iranian radar network and its status was 'rather hazy'. Furthermore, they then experienced a much stronger reaction from the Iranian air force than expected. In doubt, they opted for the 'worst case scenario' and convinced themselves that the IRIAF was not only on alert, but ready to deliver massive blows against Iraq. Correspondingly, Jaboury and his aides demanded permission to not only fly the opening air strike, but to seize and maintain the initiative by continuing to strike enemy air bases for days longer. Their conclusion was that this would prevent the IRIAF from attacking Iraqi Army units as these were exposed while moving over the open and flat terrain of south-western Khuzestan. It remains unknown if their objections were ever forwarded upwards, beyond the General Command in Baghdad; indeed, even the reaction from the latter remains unknown. One way or the other, as subsequent developments were to show, the IrAF did try to re-strike some Iranian air bases – not only after the opening strike, but after the first day of the war. It only remains unknown if such operations were undertaken at the discretion of the air force's top leadership or were sanctioned from above.[3]

UTMOST SECRECY

Planning for the Iraqi Air Force's opening strike – Operation Echo of Qadessiya – was undertaken by selected officers of the Directorate of Operations in mid-September 1980 and under conditions of utmost secrecy. Among their ideas, bomber and fighter-bomber formations were to launch at between 1134hrs and 1145hrs local time, cross the Iranian border at low altitude, and then strike a total of about 20 air bases, airports, early warning radar sites, air defence positions, and other military facilities in western and southern-western parts of the country, as listed in Table 5 (oriented from north-west towards south-east, and including details on Iranian designations and function). The first wave was timed to hit home between 1200hrs and 1230hrs (Iraqi time), when the activity of the Iranian armed forces was at its lowest, due to the noon heat. The second wave (Table 5) was to re-strike selected targets about one hour later.[4]

The secrecy surrounding Operation Echo of Qadessiya reached such proportions that the involved units were informed only on 20 and 21 September, and then in a fashion described by Sadik – then a young lieutenant serving with the Air Force Intelligence Directorate at Hurrya AB – as follows:

> At noon of 21 September, a Falcon 50 business jet – a type reserved for VIP-transportation – arrived at our air base, carrying Brigadier General Salim Sultan al-Basu. Once on the ground, he summoned the base commander, Colonel Bakir (former Su-7 and Su-20 pilot, and a veteran of wars with Israel and the Kurds), and two officers of the Intelligence Department into an office. With him, Basu brought a large white envelope with the IrAF emblem stamped in gold on the upper right side, and initials for 'top secret'. This envelope included a list of targets in Iran, available target intelligence, selected units and their formations, flight routes, and take-off times. He then issued verbal orders for execution of the operation codenamed Echo of Qadessiya. Immediately after, Colonel Bakir prepared a briefing for the commanders of the Technical Wing of al-Hurrya AB, and COs of Nos. 1, 44, and 47 Squadrons, in which he informed them about their tasks. The orders for commanders of these units were very precise, yet relatively simple: ground crews were to prepare and arm the aircraft, and then each of the units based at Hurrya was to fly the mission. Our targets were:
>
> - Tabriz AB (by Su-22Ms, which were to stage through Firnas AB, closer to the target)
> - Shahroki AB, outside Hamedan (by four Su-22s)
> - Sanandaj Airport (by six MiG-21s)
> - Saqqez Airport (by four MiG-21s)
>
> All sorties were to be flown at low altitude on the way in, and our jets were to operate in conjunction with MiG-23BNs from Abu Ubayda Ibn al-Jarrah AB.
>
> Although all the leave was cancelled, and everybody was ordered to stay at the base overnight, the involved pilots – selected on the basis of seniority – were briefed by Colonel Bakir and COs of three squadrons only at dawn of 22 September 1980.

They were informed that at noon that day, the IrAF would attack about a dozen Iranian air bases and airfields; that all formations were to be led either by wing or squadron COs; and that due to the relatively long range to their targets their aircraft would carry only two bombs each, but a maximum fuel load. The take-off was scheduled for 1130hrs Iraqi time, with all formations of the first wave crossing into Iranian airspace at 1200hrs.[5]

Despite secrecy, at least the Iraqi airmen at the Hurrya AB felt confident about their tasks because they had been running intensive exercises for exactly this kind of operation since August, and because their groundcrews had managed to make all of the available aircraft 100 percent 'fully mission capable'. Elsewhere in the air force, the reported availability of combat aircraft was at between 85 and 90 percent.[6]

TARGET INTELLIGENCE

While the quality of the strategic-level of Iraqi military intelligence on Iran in 1980 is arguable at best, it is however almost certain that the IrAF units involved in Operation Echo of Qadessiya received rather poor, and often entirely wrong, targeting intelligence. Considering the fact that a number of well-positioned Iranian defectors fled to Iraq after the failure of the coup attempt at Tactical Fighter Base 3 ('Nojeh' AB), near Hamedan, in July 1980, this is nothing but surprising. For example: while the IRIAF kept all of its combat and combat-support aircraft concentrated at its major bases – where they and their remaining crews, always suspected for disloyalty to the new government, were easier to keep under control – the GMID and other intelligence agencies not only lacked precise information about the construction of the primary enemy air bases, but assessed the IRIAF fleet of fighter-bombers as dispersed at more than a dozen different airfields in western Iran. In some cases, the Iraqis not only did not know the designations of the installations in question but lacked information even about the existence of two runways and extensive hardened aircraft shelters; in others, they considered several facilities in south-western Khuzestan as 'operational', even though their construction had been interrupted between late 1978 and February 1979, and never completed. Finally, the Iraqis were critically short not only on up-to-date maps, but indeed on precise maps of Iran. The result was unavoidable; when planning Operation Echo of Qadessiya, the General Command of the Iraqi Armed Forces, and then the High Command IrAF dispersed the effort. Moreover, they caused the leaders of several formations to miss their targets on approach, and then search for them – wasting time, fuel, and sometimes losing the moment of surprise. A review of Iraqi targets, with their Iraqi names and actual Iranian designations and purposes, is provided in Table 5.[7]

FORMATIONS AND ARMAMENT CONFIGURATIONS

The primary tactical formation of the IrAF in 1980 was a pair of aircraft. Usually, these were underway widely spaced, in a formation similar to the US Navy 'loose deuce'. This offered better mutual protection for each pilot, because the leader and wingman could – by flying line abreast – always check each other's rear. This formation also offered better concealment from visual detection, because the two widely spaced aircraft were harder to detect than two flying closely together. Iraqi interceptors rarely operated in larger formations, and – when on combat air patrol – were usually split, flying in opposite directions in their orbit, so as to cover each other's rear.

The basic formation of fighter-bombers was a widely spaced four-ship, usually flying line abreast, and with individual aircraft at slightly different altitudes (Number 1 usually took the lowest position, and Number 4 the highest). Whenever necessary, fighter-

Table 5: IrAF Operation Echo of Qadessiya, 22 September 1980; Targets	
Iraqi Designation	Iranian Designation and Function
Tabriz AB	Tactical Fighter Base 2 (Janbaz AB), IRIAF
Orumiyeh	Camp Qushchi, IRIA base
Orumiyeh	Rezaia forward operating base (FOB), IRIAF
Shahroki/Hamedan AB	Tactical Fighter Base 3 (formerly Shahroki, re-named Nojeh AB in 1979), IRIAF
Subashi	early warning radar station, IRIAF
Kabotar Ahang	Hamedan IAP, civilian facility
Kermanshah Airport	1st Combat Support Base (CSB), IRIAA
Islamabad/Shahabad Airport	Eslamabad-e-Gharb, FOB, IRIAF
Saqqez Airport	Saqqez FOB, IRIAF
Sanandaj Airport	civilian facility
Dehloran SAM site	I-HAWK SAM site, IRIAF
Dehloran	early warning radar site, IRIAF
Masjed Soleyman AB	2nd Combat Support Base, IRIAA
Dezful AB	Tactical Fighter Base 4 (Vahdati AB), IRIAF
Dezful Highway Strip	Dezful FOB, IRIAF
Andimeshk	Dezful FOB, IRIAF
Mahshahr Airport	Bandar-e Khomeyni, civilian airport & FOB, IRIAF
Agha Jari AB	Tactical Fighter Base 5 (Omidiyeh AB), IRIAF
Ahwaz IAP	civilian facility
Bushehr AB	Tactical Fighter Base 6, IRIAF
Mehrabad/Tehran AB	Tactical Air Base 1 (Mehrabad AB), IRIAF
Dowshan Tappeh/Tehran	Tactical Air Base 11, High Command IRIAF
Shiraz IAP	Tactical Air Base 7 (formerly Taddayon AB), IRIAF
Esfahan IAP	Tactical Fighter Base 8 (formerly Khatami AB), IRIAF
Bandar-e Abbas	Tactical Fighter Base 9, IRIAF

bombers could be deployed in larger formations but then would fly in a 'column of pairs', creating a chain of two, three, four or more pairs. In other cases, up to a dozen Iraqi fighter-bombers would fly in a very tight formation; in this fashion, they expected to appear as a single blip on contemporary radars, because the processors of the latter lacked the resolution, that is, the capability to distinguish between multiple aircraft underway in a tight formation. Apparently adapted from the Egyptians and Syrians of October 1973, this tactic was expected to confuse the enemy, which would expect a single aircraft, but find itself confronted by four or more.[8]

Although expecting that the IRIAF would fly CAPs down the so-called 'threat axis' – that is, between its major bases and the border with Iraq – the IrAF counted on the moment of surprise. Furthermore, the only available interceptors were short-ranged MiG-21s. Therefore, its planners decided to provide top cover for its bombers and fighter-bombers only up to the border and not inside Iranian airspace; MiG-21s were then to fly CAPs and wait for the return of attacking formations. All bomber and fighter-bomber formations were to penetrate Iranian airspace by flying along carefully selected, pre-determined routes, as low as permissible according to the terrain and the pilot's skill, and, after completing their task, withdraw at best speed in the direction of Iraq. Once they passed the Iraqi border on the way back, bombers and fighter-bombers would be 'de-loused' by IrAF interceptors flying CAPs. Exactly how many aircraft the IrAF deployed for its opening air strike remains unclear. Baghdad never released any kind of official figures, while the latest Iraqi publications citing no fewer than 192 sorties might appear 'unrealistic'; however, a quick count of those now known produces a figure of 169 (including 73 in the first wave, 53 in the second and 43 in the third).[9]

Despite placing orders for more than 200 tactical fighters in France and the USSR in the 1976–1978 period, the IrAF lacked specialised runway-penetrating and/or denial weaponry. The French-made Durandal bomb, based on Israeli experience from 1967, was on order but not yet delivered. This weapon was designed for deployment from level flight at low altitude, and equipped with a parachute, designed to slow its fall and thus enable the aircraft to reach a safe distance before a rocket motor would propel the bomb through and underneath the runway surface, where the weapon was detonated with a delayed contact fuse. The French-made BLG-66 Belouga CBU worked in a similar fashion, and this was – amongst others – to become available in a variant filled with bomblets capable of cratering runways; indeed, this highly-sophisticated weapon was developed and entered series production primarily thanks to Iraqi funding in the early 1980s. The Soviet equivalent of the Durandal was the much bigger BETAB-500, but this was still undergoing development and not yet available. Furthermore, the distance to major Iranian air bases was so large, and the endurance of most of the involved fighter-bombers so short, that their war-load had to be limited to only two weapons. Correspondingly, for their strikes on Iranian air bases, the Iraqis armed all available aircraft with only a pair of parachute-retarded FAB-500M-54ShN general-purpose high-explosive high-drag 500kg bombs, irrespective of the type of fighter-bomber in question. Belonging to the mass-produced series that entered service in 1954, the FAB-500M-54 was originally based on the M-46 series and designed for internal carriage on heavy bombers. It was foremost recognisable by a ballistic ring around its nose that acted as a vortex-generator to aid stabilisers at the rear of the bomb's body. Contrary to the M-46 series, the M-54 series included a reinforced structure making it suitable for deployment from the hardpoints of fighter-bombers at speeds of up to 1,000km/h. The FAB-400M-54 was an unguided, free-fall weapon, but compatible with all types in service with the IrAF, including MiG-21, MiG-23, Su-22, Tu-16 and Tu-22. In addition to the two bombs, the only external load carried by the involved MiG-21s, MiG-23s and Su-22s consisted of a pair of drop tanks. Configurations by type were therefore as follows:

- MiG-21s: 2 x FAB-500M-54ShN (inboard underwing pylons), 2 x 400-litre drop tanks (outboard underwing pylons)
- MiG-23BN: 2 x FAB-500M-54ShN (under-fuselage stations), 2 x 800-litre drop tanks (installed under moveable wing panels, these could be carried only with wings at full spread, and had to be jettisoned prior to attack or if the aircraft had to accelerate)
- Su-22: 2 x FAB-400M-54ShN (inboard underwing pylons), 2 x 800-litre drop tanks (main underwing pylons)

Although capable of carrying up to six FAB-500M-54s, or, alternatively, deploying much more massive weapons from the M-54 series – like the earlier-mentioned FAB-1500, FAB-3000, FAB-5000 and FAB-9000 – due to endurance-related issues, the Tu-16s and Tu-22s involved in Operation Echo of Qadessiya were armed with only three FAB-500M-54ShNs installed in their internal bomb bays.

FIRST WAVE

After taking off near-simultaneously from half-a-dozen air bases in the north and south of the country, the Iraqi bombers and fighter-bombers turned east, descended to low altitude and proceeded into Iranian airspace. In order to reach their targets – some of which were as far as 450 kilometres (280 miles) away – Iraqi MiG and Sukhoi pilots had to conserve fuel. Moreover, their jets were heavily loaded. Therefore, they flew at a relatively low speed and several formations needed up to 20 minutes to reach their targets. The first of them therefore appeared in the target zone only around 1200hrs Iraqi time (or 1300hrs Iranian time).

One of the first IrAF squadrons to reach its target was No. 5, six Su-22Ms from which were tasked with bombing TFB.2, outside Tabriz. The Squadron CO, Major Mohammed Hamid Taha, led his five wingmen into the attack at an angle of 60 degrees off the axis of the runway, at an altitude of 60 metres (197 feet) and a speed of 1,200km/h, close to the maximum limits of their FAB-500M-54. The bombs were released as planned, without any disturbance by air defences, and 11 hit the runway. However, four failed to detonate, while the 12th caused damage to a local communications facility, temporarily cutting TFB.2's connection to Tehran and Hamedan. Taha then led his formation through a 180-degree turn, and re-attacked targets of opportunity with automatic cannons (every Su-22M had two 30mm Nudelman-Rikhter NR-30 automatic guns, each with 60 rounds, installed in its wing-roots). His shells and those of his wingmen narrowly missed a Boeing 727 of Iran Air that was in the process of landing, with 143 passengers and crewmembers on board. Instead, they hit the runway behind the airliner, causing light damage to its fuselage.[10]

Air defences of the huge IRIAA base in Kermanshah were attacked by four Hunters of the FLS, which used unguided 127mm rockets to knock out a pair of 35mm Oerlikon twin-barrel anti-aircraft guns and the supporting Super Fledermaus fire-control radar. Surprisingly enough, there was no follow-up formation to exploit this success and target about 100 helicopters of the IRIAA which were parked in the open. Further south, a pair of Su-7BMKs from No. 8 Squadron bombed the IRIAF early warning radar facility at

Table 6: IrAF Operation Echo of Qadessiya, 22 September 1980; 1st Wave; Summary		
Iraqi Formation, Base	Target	Iranian Reports
Time-on-Target: 1200–1230hrs		
6 Su-22M, No. 5 Sqn, Firnas AB	TFB.2 (Tabriz)	'11 Su-22'; 11 bombs hit the runway, 4 failed to detonate; runway damaged; Iran Air B727 slightly damaged
2 Su-7BMK, No. 8 Sqn, Abu Ubayda Ibn al-Jarrah AB	Dehloran MIM-23B SAM site	'7 Su-22'; direct hit, 10 KIA
4 Hunter, FLS, Rashid AB	1st CSB (Kermanshah)	'4 Hunter'; air defence positions hit
4 MiG-23BN, No. 49 Sqn, Abu Ubayda Ibn al-Jarrah AB	TFB.3 (Hamedan)	'6 MiG-23', at 1208hrs
???	Subashi	light damage
12 Su-22, No. 44 Sqn, Hurrya AB	TFB.3 (Hamedan)	12 Su-22 at 1218hrs; runway damaged; apron damaged, control tower damaged
??? Su-22 or Su-22M, ??? Sqn	Kapotar Ahang Airport	results unknown
6 MiG-23BN, No. 49 Sqn, Abu Ubayda Ibn al-Jarrah AB	Eslamabad-e-Gharb FOB	'5 MiG-23BN' appeared at 1218hrs, attack aborted due to bad weather
4 Su-22, No. 109 Sqn, Wahda AB	Dehloran Radar Site	'2 Su-22'; radar damaged, 10 killed
5 MiG-23BN, No. 29 Sqn, Ali Ibn Abu Talib AB	TFB.4 (Dezful)	'3 MiG-23 and 2 Su-22'; 3 F-5E destroyed, 19 KIA; air defences claimed 3 MiGs as hit
4 Su-22, No. 109 Sqn, Wahda AB	TFB.4	
4 Su-22, No. 109 Sqn, Wahda AB	TFB.4	'8 Su-7'; at 1228hrs; targeted air defences
3 MiG-23BN, No. 29 Sqn, Ali Ibn Abu Talib AB	Dezful FOB	'3 MiG-23'; runway damaged
5 MiG-23BN, No. 49 Sqn, Abu Ubayda Ibn al-Jarrah AB	TFB.5	'5 MiG-23'; runway damaged
2 MiG-23BN, No. 49 Sqn, Abu Ubayda Ibn al-Jarrah AB	Ahwaz IAP	'8 IrAF aircraft'; runway damaged
4 Su-22, No. 109 Sqn, Wahda AB	Ahwaz IAP	
5 MiG-23BN, No. 29 Sqn, Ali Ibn Abu Talib AB	Masjed Suleiman AB	'5 MiG-23'; runway damaged
4 Su-22, No. 109 Sqn, Wahda AB	TFB.6	'12 Su-22', at 1226hrs; runway damaged

Dehlroan. This strike was important because around the same time multiple Iraqi formations converged in the direction of Hamedan, the local airport, TFB.3, and the Subashi early warning radar site, further east. Different Iraqi sources are in significant contradiction over who exactly hit what over the following 10 to 15 minutes – and even a cross-examination with the help of Iranian sources is not really helpful. Apparently, the first to reach the target zone was a formation of six MiG-23BNs from No. 49 Squadron, tasked with bombing TFB.3, about 60 kilometres (37 miles) north of Hamedan. At least according to the Iranians, they did so at 1208hrs, while almost simultaneously, two – perhaps more – Iraqi fighter-bombers (type unknown) hit the Subashi radar site, about 55 kilometres (34 miles) north-west of Hamedan. Ten minutes later, TFB.3 was then hit by 12 Su-22s from No. 44 Squadron, and then another Iraqi formation bombed Kapotar Ahang Airport outside Hamedan. While little is known about the results of other strikes, the dozen Sukhois targeted the runways and apron of Nojeh with a total of 24 FAB-500M-54 bombs flying straight and level, about 60 degrees off the axis, but did not re-attack with guns. While they did damage the runways – and the apron and the control tower – they failed to render this crucially important facility non-operational. Slightly more is known about the strike by five MiG-23BNs on the second CSB of the IRIAA in Masjed Suleiman, where the runway was cratered in two spots, and one hangar was completely demolished.[11]

Despite meagre results, all the pilots involved in the first wave returned to Iraq in high spirits. They had achieved complete surprise, reached nearly all of their targets entirely unmolested (except by some bad weather), deployed their bombs as intended, and came away without a single loss.[12]

SCUD INTERMEZZO

As the first wave was still underway, the Iraqis brought their ballistic missiles into the game. After receiving reports about the 'loss' of Su-20s of No. 1 Squadron, and IrAF reports about the lack of strike assets, the General Command in Baghdad ordered the 224th Missile Brigade of the Army – equipped with Soviet-made R-17E (ASCC/NATO-codename 'SS-1c Scud-B') surface-to-surface missiles – to join IrAF air strikes on TFB.4 (also known as Vahdati AB), outside Dezful, the closest main base of the IRIAF to the battlefields of south-western Khuzestan. Certainly enough, the IrAF was not to be outdone, and thus tasked three formations of MiG-23BNs and Su-22s to attack on this crucial facility. The first to approach were six MiGs from No. 29 Squadron led by that unit's CO, Major Rashid Hamad as-Sa'aydoon. One jet suffered a technical malfunction and thus only five entered Iran, and their pilots were supplied with such poor maps that they missed the target at first attempt. After burning too much fuel while circling, Sa'aydoon finally found TFB.4 and attacked. Two of his jets hit the MIM-23B I-HAWK SAM site defending the base, causing much damage and killing 18; the other

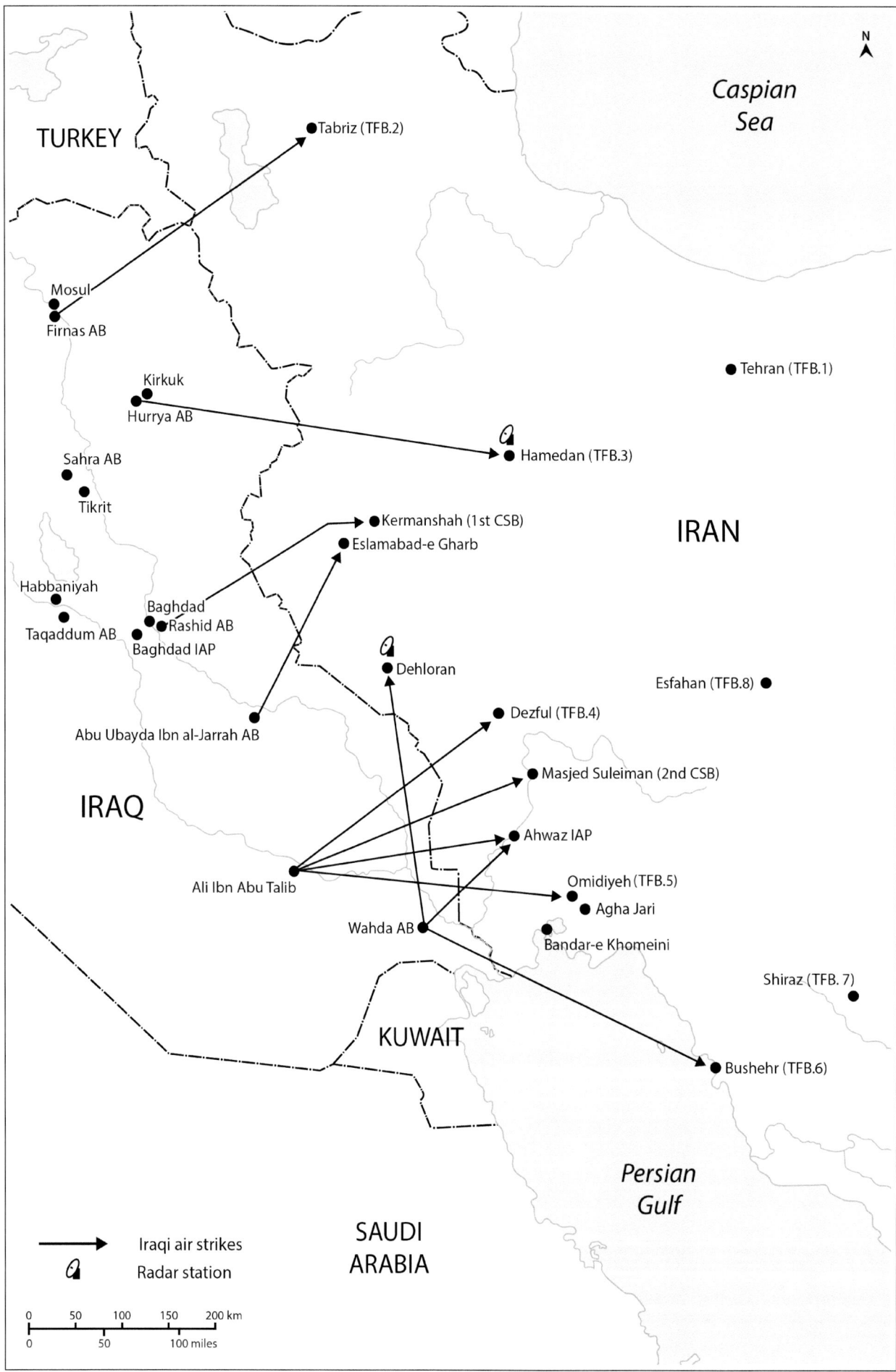

A map depicting main air strikes of the 1st Wave of Operation Echo of Qadessiya, on 22 September 1980. (Map by George Anderson)

three bombed the runway. Moreover, the entire formation then made a 180-degree turn and attacked some of the nearly 30 aircraft parked in the open, and three F-5Es were cut to pieces by 23mm shells. Although the Iranian air defences opened fire by this time and claimed three of the attacking MiGs as hit, all five jets returned safely to Ali Ibn Abu Talib AB. The stories according to which all five jets ran out of fuel and crashed – spread by some highly authoritative Iraqi sources in recent times – is at least unconfirmed, if not entirely unrealistic.[13]

Hard on the heels of the MiGs came two formations of four Su-22s from No. 109 Squadron, the pilots of which concentrated on bombing the runway. The first of them hit at 1220hrs Iraqi time, just as a group of Iranian officers was inspecting the damage caused by the first strike, and killed one of them. Finally, the 224th Missile Brigade fired its six R-17Es but these all seem to have missed by such a wide margin that nobody at TFB.4 noticed the strike. Similarly, three additional Scuds were fired by the same unit at what the GMID called the 'Andimeshk Highway Strip'; a long stretch of straight highway that could have been used as FOB in the case of emergency, north of the town with that name and about three kilometres north of TFB.4. Still, thanks to the strike led by Sa'aydoon, Vahdati AB was the heaviest hit IRIAF air base of this day, and the one that suffered most damage, and most casualties. Although the runway was repaired within a few hours, F-5Es from TFB.4 flew only CAPs for the rest of the day.[14]

SECOND WAVE

The original Iraqi plan was to re-strike selected targets about 30 minutes after the first wave. This was rather unfortunate timing because it did not offer commanders enough time to de-brief their pilots, and thus update intelligence for follow-up formations. Nevertheless, and at least according to Sadik, the second wave was launched as planned. According to Abossi, this was not the case. Instead, the High Command IrAF supposedly expressed its satisfaction with the first wave and 'replaced' the original plan with 'individual attacks' – the targets for which would have depended on the results of the first wave.[15]

Such a statement raises numerous questions, the most important of which are:
- how could Baghdad know about the results of the first wave before pilots returning to their home-bases were back on the ground and could have been de-briefed – which was a process taking longer than 30 minutes in total?
- if the plan for the second wave was abandoned, and replaced by only a few isolated attacks, then why did it include so many strikes on entirely new targets?

One way or the other, the operation went on and it included so many 'individual' strikes that the conclusion must be that the IrAF certainly flew a second wave. The sole difference was that the involved formations were smaller than those of the first wave (see Table 7 for a summary). At Firnas AB, at least

Major Rashid Hamad as-Sa'aydoon, CO of No. 29 Squadron, and leader of the most successful air strike of Operation Echo of Qadessiya. (Muthanna as-Sa'aydoon Collection)

An Iraqi pilot in the process of entering the cockpit of a MiG-23BN, seen before the war with Iran. Notable in the lower right corner is the front section of a FAB-250M-54 or FAB-500M-54 bomb; the latter was used to target the runways of Iranian air bases on 22 September 1980. (Tom Cooper Collection)

Crew chief helping the pilot of a MiG-21MF (serial number 1190) from No. 11 Squadron put on his seatbelts, at Rashid AB, in 1979. (Tom Cooper Collection)

A fully bombed-up Tu-22 rolling to take-off for the strike on Mehrabad AB on 22 September 1980. (Tom Cooper Collection)

A view towards the south of Mehrabad AB on the afternoon of 22 September 1980. Notable in the rear is a column of thick smoke, rising from the place where a C-130 Hercules transport of the IRIAF was destroyed by bombs dropped by Iraqi Tu-22s. (Tom Cooper Collection)

Another MiG-21 unit involved in the second wave was No. 47 Squadron, which flew MiG-21bis usually deployed as interceptors. Armed with two FAB-500M-54 bombs each, two formations of six jets from this unit reported achieving complete surprise and encountering no air defences at all while bombing airports at Sanandaj and Saqqez. However, a MiG-21bis was hit by the debris caused by bombs released by its formation leader during attack on Sanandaj and crashed. Its pilot, 1st Lieutenant Ra'ad Hameed, ejected safely and was captured. Considering neither airport housed any kind of IRIAF or IRIAA units, and the damage to the local runways was quickly repaired, this was actually an entirely pointless yet costly exercise.[17]

Meanwhile, what can be described as the three 'most spectacular' missions of Operation Echo of Qadessiya were developing: those including Iraqi Tu-16 and Tu-22 bombers. Sadly, no accounts from any of the involved crews are available. Instead, the flow of their operations can only be reconstructed with the help of Iranian sources. The first IrAF formation to reach its target consisted of four Tu-22 bombers, each armed with three FAB-500M-54s, tasked with striking the very heart of the IRIAF: TFB.1 – or Mehrabad AB – in southern Tehran. Flying low between the mountains of western central Iran, the bombers approached entirely undetected. The sole aircraft airborne over western Iran at the time were one Boeing 707-3J9C tanker of the Tanker & Transport Squadron, IRIAF, piloted by Major Shamsae, and a pair of F-14s on a CAP south-west of Hamedan and north of Dezful.[18] After providing fuel to the Tomcats, Shamsae turned back for Mehrabad; a few minutes later, he heard one of the F-14 pilots requesting ground control for permission to open fire. Then he heard a terse warning from the control tower of TFB.2 in Tabriz on the guard frequency, stating that the base was under attack and closed for all incoming traffic, while several fighter pilots were asking where they could land. Shamsae thus called the ground control at Mehrabad to forward such messages and issue a warning. Initially, he was not taken seriously, and it was only two minutes later that Colonel Abdolali-Zaleh, Shamsae's superior, ordered him to land his Boeing immediately, and the ground crews of his unit to prepare all available aircraft for action.[19]

a part of No. 5 Squadron was turned around – that is, refuelled and re-armed – at such a speed that it was dispatched to strike Camp Qushchi, the main base of the 64th Infantry Division of the Iranian Army. The Sukhois in turn led a formation of 12 MiG-21s from No. 11 Squadron ordered to attack the adjacent Rezaie Airport. The MiG pilots were told their target would be only two kilometres outside Orumiyeh, but this information proved wrong; nevertheless, at least a few of them did find Rezaie Airport and bombed the runway heavily, causing multiple hits – even if most of the MiG pilots, and the Su-22s leading them to the target zone, had to abort due to the bad weather. This effort was completely in vain – not only because Rezaie was a small facility without any kind of IRIAF aircraft – but No. 11 Squadron then lost one of its jets, the MiG-21MF piloted by 1st Lieutenant Saed ash-Shaykhli, which was hit while underway over a place named 'Zafraniyeh' by the Iraqis. Shaykhli managed to nurse the badly damaged aircraft all the way back to Rashid AB, but then crashed on landing and was killed. The Iraqis credited Iranian ground-based air defences with this loss, but the actual reason remains unknown.[16]

Ground crews gathering around the cockpit section of one of four Tu-22s of Nos. 18 and 36 Squadrons upon return from the air strike on Mehrabad AB, at Taqaddum, on 22 September 1980. (Tom Cooper Collection)

The latter order was not only belated, but almost had grievous consequences for the IRIAF. At 1243hrs Iraqi time, the Tu-22s appeared over Mehrabad from the southern direction. The lead aircraft released its bombs from a shallow right-hand turn, 'spraying' them all over the base. One or two hit the runway, causing only shallow 'dents' that were quickly repaired. However, the second or third bomber then released a stick of bombs on the ramp for transport units, instantly wrecking a refuelled Lockheed C-130 Hercules transport, badly damaging a Boeing 707-3J9C tanker and an empty civilian Boeing 707 airliner. Luckily for the Iranians, the resulting conflagration did not envelop any of the tankers in the process of being refuelled as per Abdolali-Zaleh's order, otherwise the IRIAF would have lost many additional aircraft. Finally, the gunner of the last Tu-22 used his twin 23mm tailgun to open fire at an F-4E Phantom II, parked in front of the Iranian Aircraft Industries workshops, pending overhaul; the precious jet was cut into two in the cockpit area. One Iranian was killed and nine wounded. On the contrary, next to nothing is known about the strike by two other Tu-22s on the Air Force Headquarters and major logistic complex at Dowshan Tappeh AB (TFB.11), in downtown Tehran, except that their bombs reportedly caused only 'minimal' damage.[20]

The request for permission to open fire from one of the IRIAF F-14 pilots heard by Major Shamsae on the radio seems to have been related to the mission of another bomber formation: four Tu-16s from No. 10 Squadron, tasked with attacking Tactical Fighter Base 8 outside Esfahan in central Iran. This strike is known to have started

Table 7: IrAF Operation Echo of Qadessiya, 22 September 1980; 2nd Wave; Summary		
Iraqi Formation, Base	Target	Iranian Reports
Time-on-Target: 1230–1300hrs		
11 MiG-23MS, No. 39 Sqn, Abu Ubayda Ibn al-Jarrah AB	1st CSB	11 MiG-23; at 1220hrs, runway badly damaged; apron hit in several spots
4 Tu-22, Nos. 18 & 36 Sqns, Taqaddum AB	TFB.1	'3 MiG-23' or '3 Tu-22', at 1243hrs; 1 C-130 destroyed; 1 B707 written off; 1 F-4E badly damaged (repaired after the war)
2 Tu-22, Nos. 18 & 36 Sqns, Taqaddum AB		
6 MiG-23BN, No. 49 Sqn, Abu Ubayda Ibn al-Jarrah AB	TFB.3	'4 Su-22, 2 Tu-22'; at 1240hrs; runway badly damaged; airport shut down
4 Tu-16, No. 10 Sqn; Taqaddum AB	TFB.8	'3 Tu-16', at 1251hrs; runway and some shelters lightly damaged; 1 Tu-16 crashed
6 MiG-21, No. 47 Sqn, Hurrya AB	Sanandaj	'16 MiG-21'; runway badly damaged; 1 MiG-21 crashed
6 MiG-21, No. 47 Sqn, Hurrya AB	Saqqez	results unknown
6 Su-22M, No. 5 Sqn, Firnas AB	Camp Qushchi & Rezaie Airport	'6 MiG-23'; scheduled for 1310hrs; aborted due to bad weather
12 MiG-21MF, No. 11 Sqn, Rashid AB	Rezaie Airport	aborted due to bad weather; 1 MiG-21 crashed
4 Su-22, No. 109 Sqn, Wahda AB	Ahwaz IAP	'1 MiG-23, 4 MiG-21', at 1330hrs; attacked radar system and other facilities; results unclear
4 Su-22, No. 109 Sqn, Wahda AB	Mahshahr Airport	results unknown

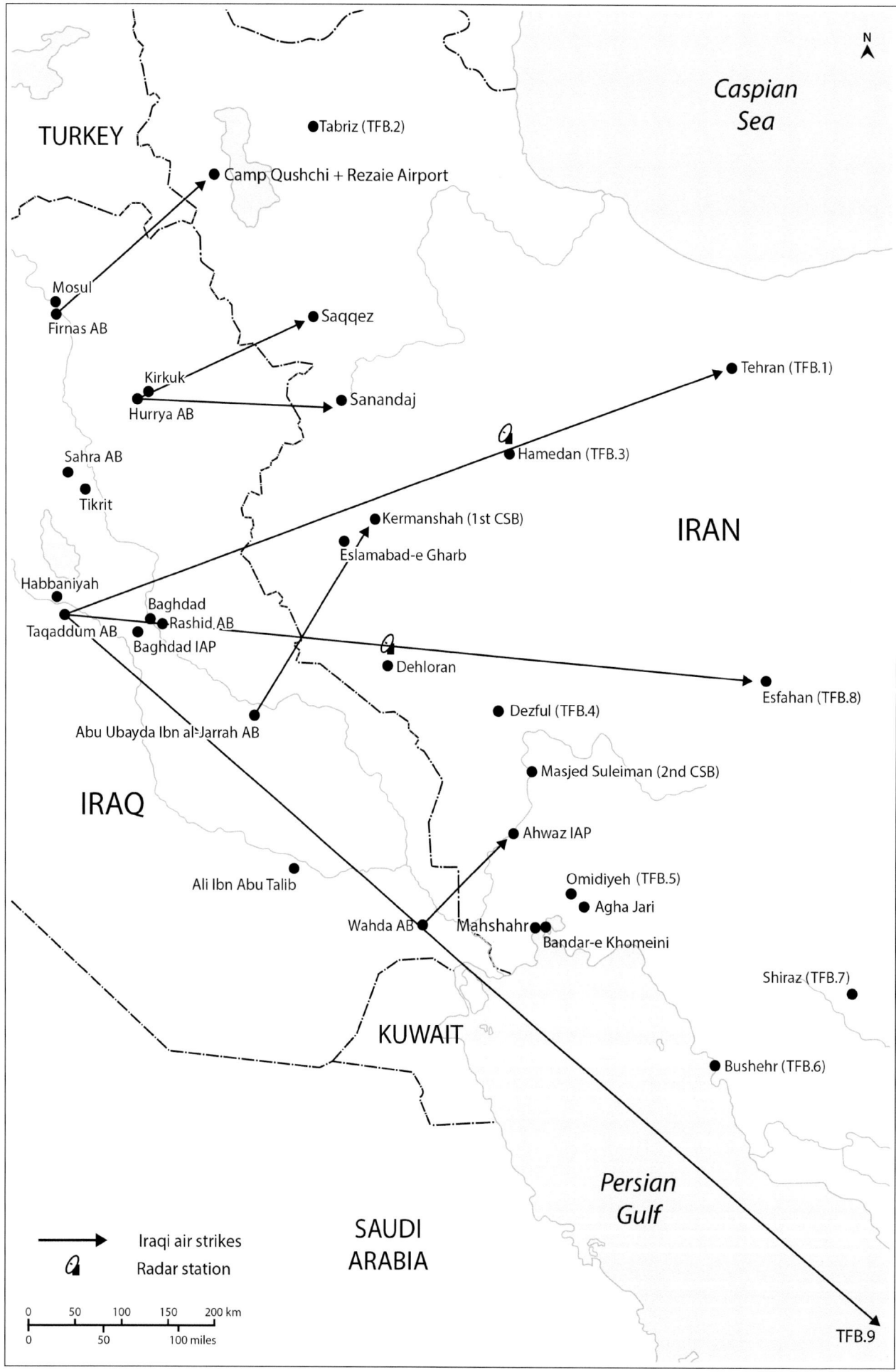

A map depicting the main air strikes of the second wave of Operation Echo of Qadessiya on 22 September 1980. (Map by George Anderson)

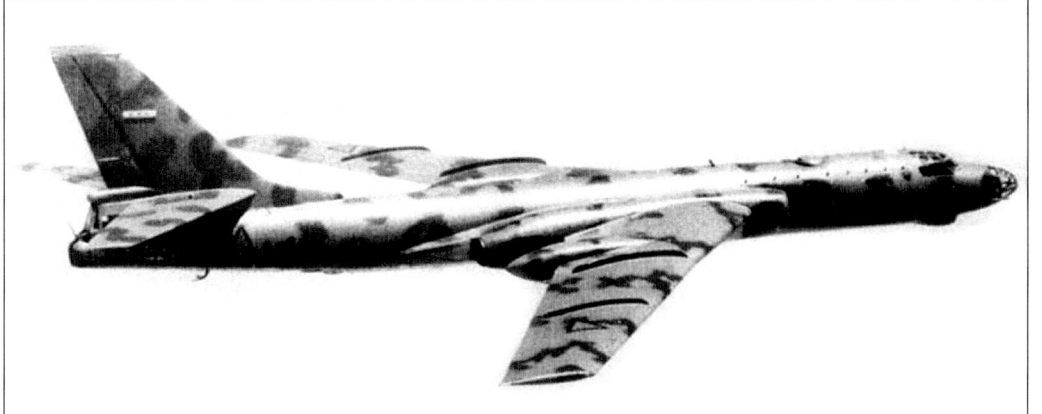

In 1973, around the time that the IrAF started receiving Tu-22 bombers, all the surviving Tu-16s of No. 10 Squadron received a camouflage pattern in grey-green and olive green on their top surfaces and sides. This Tu-16 was photographed in late 1973 by interceptors of the US Navy. (US Navy Photo)

Wreckage of the Tu-16 piloted by Lieutenant Colonel Adel Abdul Hamid Uthman, as found by the Iranians on 23 September 1980. Uthman was the highest-ranking officer of the IrAF killed not only during Operation Echo of Qadessiya, but during the entire war with Iran. (Tom Cooper Collection)

falling apart as the bombers entered the Summar Hills in western central Iran on the way to their target. According to the Iraqis, the lead aircraft, piloted by CO No. 10 Squadron, Lieutenant Colonel Adel Abdul Hamid Uthman, with Major Muhannad al-Awsi as navigator, crashed into a mountain killing everybody on board, including not only Uthman and Awsi, but also Major Azhar Abd al-Karem (co-pilot), 1st Lieutenant Abbas Qatran (radio-operator/navigator), Warrant Officer Anwar Razzuqy (gunner), and Warrant Officer Khalid Abd ar-Razzaq (gunner). The crews of the surviving three aircraft did attempt to press their attack home but made a navigational mistake and ended up releasing their bombs at the steel works of Esfahan, with unknown results. All three returned safely to Taqaddum AB.[21]

The final known air strike from this wave hit the construction site of Mahshahr airport, near Bandar-e Khomeini, a giant port developed in the late 1970s within the swamps on the northern verge of the Persian Gulf. Sadly, only elementary details about the involved Iraqi formation are known – as listed in Table 7.[22]

THIRD WAVE

Meanwhile it is certain that additional Iraqi air strikes were flown later in the afternoon, in effect creating a third wave (summarised in Table 8). There is very little information about most of the missions in question, just as it remains unknown if these were launched at the discretion of the High Command IrAF – or on orders from higher up. From what can be made out, the third wave had two high points. The first was in the north; as soon as all 12 Su-22s of No. 44 Squadron were back to Hurrya AB from their attack on TFB.3, they were refuelled and re-armed, and ordered to strike the same target again. For unknown reasons, only six jets flew this mission. Nevertheless, they caused additional damage to the runway – disrupting repair works there – and to one of the underground fuel depots.[23]

A mission actually launched during the second wave, but due to the distance to its target it actually took place during what became the third wave, and this became probably the best example of what happens when incompetent and unaccountable politicians start micromanaging a military operation. Encouraged by the positive reactions from the Kuwaiti and Saudi governments for his idea for an invasion of Iran, Saddam Hussein ordered an air strike on TFB.9, outside Bandar-e Abbas – a place in the middle of the Hormuz Straits and thus hopelessly outside the range of even the longest-ranged bombers in IrAF arsenal. Orders were orders, and thus although the IrAF was critically short on strike assets, and the aircraft in question could have been better deployed somewhere else, the mission was assigned to two crews drawn from Nos. 18 and 36 Squadrons. They flew their Tu-22s down the Persian Gulf, across Qatar and the United Arab Emirates, and all the way to Seeb Air Base in Oman, intending to refuel there and then fly the actual attack. However, nobody in Baghdad thought of at least informing the Omani authorities about this operation, not to mention actually securing their agreement and support. Therefore, on arrival in Seeb the two Tu-22 crews were badly disappointed – indeed, frustrated – to find out that all the personnel were British. This included the base commander who flatly refused to refuel their aircraft. Undaunted, the leader of the Iraqi formation continued insisting, demanding that the 'Omanis' help him 'strike at the common enemy'. Unsurprisingly, the situation escalated to a point where not few a expletives were exchanged. Ultimately, the British did agree to provide only enough fuel for the two Tu-22s to return back to Iraq via Saudi Arabia, advice eventually followed by the two Iraqi crews.[24]

The final known mission of Operation Echo of Qadessiya took place around 1700hrs, shortly before sunset. It included four Hunters from the FLS, piloted by four highly experienced instructor pilots: Major Wallid Younis, Major Abdul Qadir as-Sabbah, Major Zuhair Abd al-Hassoun and Major Riyad ar-Rawi. On orders from above, they launched to re-attack the airport outside Eslamabad-e-Gharb – actually an unimportant place serving as a forward operations

Table 8: IrAF Operation Echo of Qadessiya, 22 September 1980; 3rd Wave; Summary		
Iraqi Formation, Base	Target	Iranian Reports
Time-on-Target: 1650–1710hrs		
4 Su-22, No. 109 Sqn, Wahda AB	Agha Jari	'5 MiG-23', at 1650hrs; bombed runway and four hardened aircraft shelters, no damage reported
7 MiG-23MS, No. 39 Sqn, Abu Ubayda Ibn al-Jarrah AB	1st CSB	'7 MiG-23', at 1652hrs–1655hrs; bombed runway and air defence positions north-east of 1st CSB
6 MiG-23MS, No. 39 Sqn, Abu Ubayda Ibn al-Jarrah AB	1st CSB	as above
6 Su-22M, No. 44 Sqn, Hurrya AB	TFB.3	'4 Su-22', at 1621hrs, followed by '2 Su-22' at 1625hrs, hit the main runway disrupting repair works, and fuel depot
4 Hunter, FLS, Rashid AB	Eslamabad-e-Gharb FOB	results unknown; 1 Hunter shot down
9 MiG-23BN, No. 29 Sqn, Ali Ibn Abu Talib AB	TFB.4	'6 MiG-23', at 1710hrs, runway and taxiway damaged
5 MiG-23BN, No. 49 Sqn, Abu Ubayda Ibn al-Jarrah AB	unclear	unknown
2 Tu-22, Nos. 18 & 36 Sqns; Taqaddum AB	TFB.9	mission aborted after landing at Seeb AB in Oman

Major Abdul Qadir Qadir as-Sabbah was a highly experienced fighter-weapons instructor serving with the FLS: he went missing in action at sunset of 22 September 1980 while flying an entirely pointless air strike against a secondary airfield outside Eslamabad-e-Gharb in western Iran. (via Ali Tobchi)

base for the IRIAF only in the case of emergency, hardly worth much attention – and then so late that due to the darkness and the requirement for radio silence, the Numbers 3 and 4 of the formation failed to join the first two jets and aborted the mission. Worse yet, by now the Iranians were on alert and their anti-aircraft gunners were in position and keen to engage whatever flew by. Unsurprisingly, the remaining two Hunters encountered fierce resistance and the jet flown by Major Sabbah was hit by ZU-23 anti-aircraft guns. The pilot managed to turn his aircraft back in the direction of Iraq and reached an area close to the border before being forced to eject; he was never seen or heard of again. The IrAF thus lost another precious fighter-bomber, and a highly experienced and promising pilot.[25]

RESULTS

As mentioned above, Iraqi sources claimed that Operation Echo of Qadessiya was a major success, and that their strikes not only took the Iranians by surprise, but even caused shock and chaos, and hit the IRIAF very heavily, too. While not only the IRIAF, but indeed the entire Iranian armed forces were in a state of chaos since earlier, and while the Iraqis certainly achieved surprise, the actual results of this first blow can only be assessed as 'poor'. The Iraqis did manage to damage runways on three major Iranian air bases – TFB.2, TFB.3, and TFB. 4 – but they knocked out only a handful of aircraft: three F-5Es of the IRIAF and one Boeing 707 airliner of the IRIA, and damaged about a dozen others. The fact was that the US and Israeli-designed air bases of Iran were at least as heavily fortified as the Yugoslav-designed and constructed air bases in Iraq. Alone this made it unlikely that any other contemporary air force in the Middle East would have achieved significantly different results in an attack on IRIAF air bases. Combined with the lack of better targeting intelligence, and the lack of runway-cratering and runway-denial munition, this resulted in the Iraqis being in a position where they were unlikely to achieve more than they did, and that is true regardless of whether the IrAF flew 192 strike sorties, as claimed by most recent Iraqi sources, or more than 200, as the above-listed cross-examination of Iraqi and Iranian sources indicates. In turn, the fact that its opening blow did catch the Iranians by surprise, did help the IrAF to avoid suffering heavy losses; in total, it had to write off two MiG-21s, a Hunter, a Tu-16 and at least three highly experienced fliers.

IRANIAN REACTION

Having had most of its top commanders arrested, sentenced to death and executed already by spring 1979, shortly after Khomeini's take over, and the mass of its mid-ranking commanders then successively purged through 1980 and replaced by under-qualified officers, chaos still reigned within the chain of command of the IRIAF, as well as the entire Iranian armed forces. Out of the nearly 100,000 personnel of two years before, almost two-thirds had either been purged, or had left the service on their own. Except for units flying F-5E/Fs – all of which were concentrated at TFB.2 and TFB.4 – all other wings and squadrons were critically short of qualified pilots, radar intercept officers, and ground personnel. Nevertheless, those that were still around went into action almost immediately, instantly proving that the aim of Saddam's leadership to suppress IRIAF activity for the first 48 hours of the war by the opening air strike, resulted in a failure.

The first to go was the CO of the TFB.3-based 32nd Tactical Fighter Squadron (TFS), Major Jalil Pour-Rezaee. Taking off at 1600hrs (1500hrs Iraqi time), he led four F-4Es of the Alpha Red Flight armed with five 170kg (375lbs) M117 general purpose bombs, 170kg each, in an attack on Abu Ubayda Ibn al-Jarrah AB, outside al-Qut. Flying extremely low, the four Phantoms avoided detection before jinking upwards for their terminal attacks; however, the Iraqi anti-aircraft gunners were on alert and promptly opened fire. While the lead F-4E bombed its target, the Number 2 of the Iranian formation was hit midway through its terminal dive, and went straight down, killing the crew. The other two Phantoms completed their mission. Even if the damage caused by this air strike was quickly repaired by the Iraqis, the bombs dropped by Alpha Red Flight were the first Iranian weapons to actually hit Iraqi soil.[26]

Only minutes later, Captain Sepidomooy-Azar led four F-4Es armed with two 454kg (1,000lbs) Mk.83 bombs from the TFB.6-based 61st TFS into an attack on Wahda AB. Streaking very low along the coast of the Persian Gulf, via Bandar-e Deilam and Qorveh, to Khosro-Abad, and across the Shatt al-Arab north of Umm al-Qasr, they did manage to catch the Iraqis by surprise. According to Iraqi reports, all their bombs missed the intended target – the runway – but according to US reports, they did wipe out the military music band of the base, killing 28 out of 34.[27]

The two Iranian air raids flown in reaction to the Iraqi actions might appear little in the grand total; indeed, they might be misunderstood as confirming the effectiveness of the opening IrAF blow. However, any such illusions were to be proven completely wrong within less than 24 hours.

CONCLUSIONS

Between 1970 and 1980, the Iraqi Air Force went through a high-paced and turbulent period, in the course of which it experienced a fundamental transformation from a third-class 'royal flight club', into a seasoned military service with the potential of a war-decisive branch. Certainly enough, by 1980 very little of the high technology ordered since 1976 was already in the country, and this transformation was anything but obvious. However, perhaps the biggest change the IrAF experienced during this period of time was the one within. Most important was the change of mood of operation within its top ranks. Like Prussian and German officers of the nineteenth and the early twentieth century who were obsessed with the idea of winning wars by re-fighting the Battle of Cannae – through encircling and then destroying the enemy in the form of what they called the 'Kesselschlacht' – so the Iraqi military officers of the same period, and well into the 1960s, were obsessed with the idea that only they knew what was best for the country, only they were right, and that therefore they had the duty of not only meddling in politics, but indeed of running and building up the nation. Initially, the officers in question all had Ottoman military backgrounds but by 1958 that generation was completely swept aside by pan-Arabists, nationalists, republicans and leftists, resulting in as many as two dozen military coups and coup attempts. Ironically, it took one of the weakest political parties of Iraq taking power in Baghdad, and brutal persecution by 'security' services developed by Saddam Hussein ostensibly in the name of the Ba'ath Party (though actually in the interests of establishing him in power), to bring the armed forces under control. Another irony was that this became possible through Major General Ahmed Hassan al-Bakr – a military officer – siding with the civilian wing of the Ba'ath, and then at the cost of merciless purges of all the possible competitors, including several of Iraq's top airmen. Bakr then led the country into a period of its greatest political stability and greatest economic development ever, while all the time giving up to Saddam Hussein. Yet another irony is that it was this combination at the top in Baghdad that resulted in the Iraqi armed forces – and so also the IrAF – experiencing a period of their most rapid growth, and high professionalisation within the officer corps – and this *despite* Saddam's policy of advancing and appointing his favourites and supporters. For a while, and after more than 40 years of existence, the IrAF was reformed into a highly professional and coherent military service, about to become equipped with some of the best technology available. The final irony is that this system functioned only as long as Saddam was still under at least the nominal control of Bakr: almost as soon as the latter was forced into retirement, the new strongman in Baghdad led the country straight into a war with the Islamic Republic of Iran, primarily due to completely irrational ideas about himself as a successor to the great Babylonian conquerors, a descendant of the Prophet Mohammed, and a new leader of the Arab world. Unsurprisingly, within a few weeks, all of the professionalism that the armed forces had achieved at immense cost during the late 1970s was swept aside, and – despite much enthusiasm and serious effort – the politicians manoeuvred the air force straight into dilettantism: it was on orders from Saddam Hussein and the RCC of the Ba'ath Party, that the Iraqi armed forces then launched the invasion of Iran on 22 September 1980 – under conditions that can only be described as ironic. On one hand, their officers and other ranks were indoctrinated to think that the Iranians were in possession of superior forces and were poised to invade Iraq; on the other, they had been told to expect a quick and easy victory against an enemy certain to run on sight. By this time, the power of Saddam Hussein and his supporters, and thus the level of control they exercised over the Iraqi armed forces, was absolute. If anybody in Iraq decided to question this contradiction, he or she had to remain silent in order to survive. Nevertheless, the invasion was actually greeted with much enthusiasm and chest-thumping – in Baghdad, in Basra, in Mosul and in the squadron ready rooms of the IrAF – and this was further bolstered by an official announcement that urged the 'Arab population of Iraq' to 're-create the Battle of Qadessiya' – the clash from the year 637 when Moslems of the Arabian Peninsula defeated Sassanid Persia and proceeded to capture Ctesiphon, the Sassanid capital on the Tigris River, near contemporary Baghdad and then to expel the Persians from all of what is modern-day Iraq. How the members of the various minorities that served with the IrAF and other branches at the time have felt about such statements remains unrecorded.

Finally, and with hindsight, it cannot but be concluded that nobody in Iraq – from Saddam Hussein to the last private on the ground – had any kind of a clear strategic goal or aim, nor a trace of an idea of how long the war with Iran would last; certainly enough, the intention was for the army to drive the Iranian artillery away from Iraqi borders, and the air force was to keep the IRIAF at bay. However, all the while, it remained entirely unclear exactly how Iran – a country with three times bigger a population and three times larger total area than Iraq – was to be brought to its knees. There was no strategic concept, only a very limited operational and tactical focus, and very few ideas about strategic and operational aims and implications.

Much to the surprise of the Iraqis, instead of running away, united by the challenge of an aggressor on their soil, the Iranians were to put up fanatical resistance – and this resistance was to get ever more bitter, the deeper the Iraqis attempted to advance, and the longer the war went on. The 'power of Iranian nationalism' was a factor that not only Saddam, but all of his aides and supporters had massively underestimated when expecting their enemy to quickly fold.

From that point of view, it might appear as if it was irrelevant if the strongman in Baghdad and his aides micromanaged the IrAF into delivering a largely ineffective first blow or sent the army to trundle into Khuzestan while lacking even up-to-date maps, but in the hope that something might turn up. However, as the next volume of this mini-series will show, there is little doubt that had the IrAF been left to operate professionally, or had Saddam only waited for a more opportune moment to attack, at least the Iraqi Air Force would have had a serious chance of preventing the most grievous catastrophe that ever befell Iraq.

Table 9: Known Serial Numbers of IrAF Aircraft, 1970–1980[28]		
Serial	Aircraft Type	Notes
906-913	MiG-15bis/MiG-15UTI	7 delivered 1970 (from Czechoslovakia)
914-915	???	
916-940	Zlin 526FI	25 del. 1971
941-955	Zlin 526AF	15 del. 1971
956-957	MiG-21US	del. 1971; refurbished by Flugzeugwerke Dresden, 1980s
966-???	MiG-15bis/MiG-15UTI	10 del. 1971 (from Czechoslovakia)
978-1000	Su-7BMK	23 planned for 1970, del. not confirmed
1002-1006	MiG-21M	del. 1970: refurbished by Flugzeugwerke Dresden, 1982–83
1009	MiG-21MF	del. 1971; refurbished by Flugzeugwerke Dresden, 1980s
1019	MiG-21MF	del. 1972; 2 kill markings (see colour section)
1019-1034	MiG-15bis SB	16 del. 1971 (from Czechoslovakia)
1036-1038	MiG-21UM	del. 1971–72; 1036 refurbished by Flugzeugwerke Dresden
	MiG-15bis	6 del.1972 (from Czechoslovakia)
1041	MiG-23MS	
1051-1052	MiG-21MF	2 del. 1973
1076	SA-316 Alouette III	16 del. 1971–72
1093-1097	MiG-19S	12-18 del. 1972
1098	Mi-6	16 del 1972–73
1099	MiG-21MF	del. 1973
1101	Mi-6	
1109-1122	Tu-22A/U	14 del. Oct 73
1104-1105	MiG-17F/PF	del. mid-1973
1109-1122	Tu-22A/U	14 del. Oct 73
1123-1160	L-29 Delfin	38 del. Dec 73–Apr 74
1162-1173	Su-20	del in 1973
1181-1195	MiG-21MF	del 1973–1974; refurbished by Flugzeugwerke Dresden, 1980s
1214	SA-316 Alouette III	
1227-1230	Su-7BMK	4 planned in mid-1970s, del. not confirmed
1232	MiG-21MF	Delivered in mid-1970s
1254-1260	L-39C	6 delivered in May 1975, first delivery batch
1256	SA.342M	
1268-1269	Mi-8T	
1269-1273	L-39ZO	
1274-1276	MiG-23UB	140 ordered in Feb 72; c/n of 1275 was 1037408
1276-1279	L-39ZO	15 del. Nov 75–Feb 76
1295-1315	Su-20	del. 1973–1975

Table 9: Known Serial Numbers of IrAF Aircraft, 1970–1980[28]		
1340-1368	L-39ZO	27 del. Oct 76–Jun 77; 28th crashed before delivery
1369-1375	L-39ZO	
1374	Su-20	del. 1976
1397-1398	SA.342M	
1422	SA.330L	
1428	SA.342M	
1428	MiG-23BN	
1429	SA.342M	
1446	SA.342M	
1449	MiG-23MS	
1455	MiG-21MF	refurbished by Flugzeugwerke Dresden, 1980s
1457	MiG-21MF	refurbished by Flugzeugwerke Dresden, 1980s
1460-1461	SA.342M	
1463-1471	MiG-21UM	12 del. 1975–76
1472-1573	SA.342M	
1540	MiG-23MF	5 del. 1980
1544	MiG-23MF	
1574	Su-22M	
1574-1592	SA.342M	
1605-1607	SA.342M	
1618	MiG-23BN	
1625-1688	SA.342M	
1674-1675	MiG-23UB	
1702	Su-22UM	
1736	Su-22	
1821	SA.342K	
1823-1827	SA.342K	
1828	MiG-21R	possible attrition replacement
1829-1847	SA.342K	
1836	Su-22	
1898-1899	SA.342K	
1907-1908	SA.342K	
1916-1917	SA.342K	
1925-1926	SA.342K	
1934-1935	SA.342K	
1943-1944	SA.342K	
1952-1953	SA.342K	
1960-1961	SA.342K	
1969-1970	SA.342K	
1980-1981	SA.342K	
2003-2005	SE.316C	
2017	SE.316C	
2020-2027	SA.321GV	12 del. Aug 76–Oct 77
2021	Su-22	
2022-2023	SE.316C	
2026	Su-22M	
2031-2031	SE.316C	

Table 9: Known Serial Numbers of IrAF Aircraft, 1970–1980[28]		
2039-2040	SE.316C	
2040	SE.316C	
2043-2044	SE.316C	
2050-2051	Su-20	2050 re-serialled as 20503 in 1989
2052	SE.316C	
2055-2056	SE.316C	
2062	Mi-8T	
2062-2063	SE.316C	
2070-2071	SE.315C	
2074-2079	Su-22	2076 re-serialled as 22566 in 1989
2080-2081	SE.316C	
2086-2088	SE.316C	
2092-2096	SE.316C	
2101-2102	MiG-21bis	
2103-2105	SE.316C	
2110	Mi-25	
2118-2119	Mi-25	
2120-2122	SE.316C	
2122	MiG-21UM	
2126-2127	SE.316C	
2128	Mi-25	
2133-2134	SE.316C	
2142	An-2B	
2157	Su-22M	
2170-2171	SE.316C	
2175-2176	SE.E16C	
2181-2183	SE.316C	
2188-2189	SE.316C	
2194-2195	SE.316C	
2250	Su-22	
2255	Su-22M	

REFERENCES

INTERVIEWS

Major General Alwan al-Abossi (IrAF), July 2014
Major General Qaldoon Q. Bakir (IrAF), March 2007
Major General Hicham Barbouti (IrAF), August 2018
Major General Muwaffak Saeed Abdullah an-Naimi (IrAF), February 2007
Major General Abdel Moneim Zaki Okasha (EAF), July 2011
Brigadier General Ahmad Sadik Rushdi al-Astrabadi (IrAF), March 2005, March 2006, March 2007 and October 2007
Air Commodore Fikry el-Gindy (EAF), February 1999
Captain Muthanna as-Sa'aydoon (IrAF), October 2021 (son of Major Rashid Hamad as-Sa'aydoonm, CO No. 29 Squadron, IrAF, as of September 1980; and an active officer of the Iraqi Air Force)

PRIMARY DOCUMENTS

Written excerpts and photographs from the private documentation of the following veteran IrAF pilots were provided by Brigadier General Ahmad Sadik:
- Mohammad Ahmad
- Fayez Baqir
- Rabee' Dulaymi
- Major General Haytham Khattab Omar ('Memoirs of the Commander of the Iraqi Air Force', privately published document from 2002)
- Ahmad Sadik, *Iraqi Air Force Electronic Warfare during the Iran-Iraq War, 1980–1988* (unpublished document from 2007)
- Jameel Salwan
- Mohammad Salman
- Mohammad Saaydon ('Pilot Memoir', privately published document from 2005)

Air Ministry (UK), files as listed in endnotes
British Defence Attaché in Iraq (UK), files as listed in endnotes
CIA, *Report Z-20116/80*, September 1980
CIA, *Report IAR-0254/80*, September 1980
CIA, *Iraq's Air Force: Improving Capabilities, Ineffective Strategy; An Intelligence Assessment*, October 1987, CIA Freedom of Information Act, Electronic Reading Room
DIA, *Electronic Warfare Forces Study – Iraq*, 9 August 1990, National Archives
Dossier on the Role of the Iraqi Air Force in the Gulf War', by the US Department of Defense-sponsored Conflict Records Research Center (CRRC Record Number SH-AADF-D-000-396) in the course of 'Project Harmony')
Foreign Office (UK), files as listed in endnotes
Foreign Technologies Division (USAF), *Fishbed C/E Aerial Tactics* (tactical manual for MiG-21F-13, MiG-21PF, MiG-21FL and MiG-21PFM, obtained from Iraq in 1963 and translated to English by the Foreign Technologies Division USAF, in 1964)
Iraqi Air Force & Air Defence Command, *An Analytical Study on the Causes of Iraqi Aircraft Attrition During the Iran-Iraq War* (in Arabic), (self-published for internal use, May 1991; English transcription provided by Sadik)
Iraqi Air Force & Air Defence Command, *Analytical Study of Iraqi Aircraft Attrition During the Iran-Iraq War* (in Arabic), (self-published for internal use, September 1991; translation provided by Brigadier General Ahmad Sadik)
Iraqi Air Force & Air Defence Command, *The Role of the Air Force and Air Defence in the Mother of All Battles: After Action Report* (in Arabic), (self-published for internal use, 5 November 1991, captured in 2003 and translated as 'A 1991 Dossier on the Role of the Iraqi Air Force in the Gulf War', by the US Department of Defense-sponsored Conflict Records Research Center (CRRC Record Number SH-AADF-D-000-396) in the course of 'Project Harmony')
Iraqi Air Force & Air Defence Command, *Engine-Related Problems with Su-20s, 4 September – 24 October 1980* (in Arabic), (self-published for internal use, 29 May 1981; translation provided by Brigadier General Ahmad Sadik)
Iraqi Air Force Martyrs Website, 1931–2003, *iraqiairforcememorial.com*
Iraqi Ministry of Foreign Affairs, *Iraqi Air Force Aircraft flown to Iran, in 1991* (in Arabic), (letter to the General Secretary of the UN, September 1991)
IRIAF, *204 KIA and 58 POW pilots of the Sacred Defence* (in Farsi), (self-published for internal use, listing 204 IRIAF pilots and crewmembers that were killed and 58 that were captured during the war with Iraq by their full rank and name, aircraft they flew, date of their death or captivity; date and place of issue unknown; copy provided by Farzin Nadimi)
Ministry of Defence (UK), files as listed in endnotes
Office of Naval Intelligence (USA), files as listed in endnotes

BIBLIOGRAPHY

Abossi, Major General A., *In Memory of the Comprehensive Response* (Amman: self-published document, September 2010)
Alnasrawi, A., *The Economy of Iraq: Oil, Wars, Destruction of Development and Prospects, 1950–2010* (Westport: Greenwood Press, 1994)
Awardi, A. al-, *History of the Iraqi Armed Forces, Part 17: Establishment of the Iraqi Air Forces and its Development* (in Arabic), (Baghdad: Ministry of Defence, 1988; translation by Ali Tobchi)
Bergquist, Maj R. E., *The Role of Airpower in the Iran-Iraq War* (Maxwell AFB: Air University Press, 1988)
Boutz, G. M. & Williams, K. H., *U. S. Relations with Iraq: From the Mandate to Operation Iraqi Freedom* (Air Force History and Museums Program: Washington D.C., 2015)
Collier's Encyclopedia Yearbook 1979 – Covering the Year 1978 (New York: Crowell Collier and Macmillan, 1979)
Collier's Encyclopedia Yearbook 1980 – Covering the Year 1979 (New York: Crowell Collier and Macmillan, 1980)
Cooper, T. & Sandler, E., *Lebanese Civil War, Volume 2: Quiet before Storm, 1978-1981* (Warwick: Helion & Co., 2021)
Cooper, T. & Sipos, M., *Iraqi Mirages: The Dassault Mirage Family in Service with the Iraqi Air Force, 1981–1988* (Warwick: Helion & Co., 2019)
Cooper, T., *MiG-23 Flogger in the Middle East: Mikoyan I Gurevich MiG-23 in Service in Algeria, Egypt, Iraq, Libya and Syria, 1973-2018* (Warwick: Helion & Co., 2018)
Cooper, T. & Nicolle, Dr. D., *MiGs in the Middle East, Volume 2: Soviet-designed Combat Aircraft in Service in Egypt and Syria, 1963–1967* (Warwick: Helion & Co., 2021)
Cooper, T., Sadik, Général de Brigade A., Bishop, F., *La guerre Iran-Irak: Les combat aériens, Hors-Serie Avions No. 22 & No. 23* (Outreau: Éditions LELA PRESSE, 2007)

Dildy, Douglas C. & Cooper, T., *F-15C Eagle vs MiG-23/25, Iraq 1991* (Oxford: Osprey Publishing, 2016)

Dupuy, Col T. N. & Blanchard, Col W., *The Almanac of World Military Power* (2nd Edition), (London: Arthur Barker Ltd, 1972)

Flintham, V., *Air Wars and Aircraft: A Detailed Record of Air Combat 1945 to the Present* (London: Arms and Armour Press, 1989)

Francona, R., *Ally to Adversary: An Eyewitness Account of Iraq's Fall from Grace* (Annapolis: Naval Institute Press, 1999)

Garello, G., 'Les Ailes Italiennes en Iraq (1937-1941)', *Avions,* No. 183 (September–October 2011)

Group of Authors, *The Role of the Iraqi Armed Forces in the October 1973 War* (in Arabic), (Beirut: Establishment for Arab Studies and Publication, 1974)

Fukuyama, F., *The Soviet Union and Iraq since 1968* (Santa Monica: RAND, 1979)

Hiro, D., *The Longest War: The Iran-Iraq Military Conflict* (Routledge: Chapman and Hall Inc., 1991)

Hooton, E. R., Cooper, T. & Nadimi, F., *The Iran-Iraq War, Volume 1: The Battle for Khuzestan, September 1980–May 1982* (Revised Edition), (Warwick: Helion & Co., 2019)

Hooton, E. R., Cooper, T. & Nadimi, F., *The Iran-Iraq War, Volume 2: Iran Strikes Back, June 1982 – December 1986* (Revised Edition), (Warwick: Helion & Co. 2019)

Hooton, E., R., Cooper, T. & Nadimi, F., *The Iran-Iraq War, Volume 3: Iraq's Triumph* (Solihull: Helion & Co., 2017)

Hooton, E., R., Cooper, T. & Nadimi, F., *The Iran-Iraq War, Volume 4: The Forgotten Fronts* (Solihull: Helion & Co., 2018)

Hoyt, T. D., *Military Industry and Regional Defense Policy; India, Iraq and Israel* (Oxon: Routledge, 2007)

Khalil, S. al-, *Republic of Fear: The Inside Story of Saddam's Iraq* (London: Hutchinson Radius, 1989)

Liébert, M. & Buyck, S., *Le Mirage F1 et les Mirage de seconde generation à voilure en flèche, Vol.1* (Outreau: Éditions LELA PRESSE, 2007)

Marashi, I. al- & Salama, S., *Iraq's Armed Forces: An Analytical History* (Abingdon: Routledge Middle Eastern Military Studies, 2008)

Metz, H. C., *Iraq: A Country Study* (Washington: GPO for the Library of Congress, 1988)

Murray, W. & Woods, K. M., *The Iran-Iraq War: A Military and Strategic History* (Cambridge: Cambridge University Press, 2014)

Nadimi, F., *The Iranian Oil Industry and The Iran-Iraq War of 1980–88* (PhD thesis, The University of Manchester, 2011)

Namaki, Brigadier General A. R., Khalili, Brigadier General H., Djavaheri, Colonel. A. R., *History of Air Battles, Volume 2: Invasion and First Response* (in Farsi), (Tehran: Centre of Strategic Studies, IRIAF, 2016)

Navias, S. & Hooton, E. R., *Tanker Wars: The Assault On Merchant Shipping During The Iran-Iraq Conflict, 1980–1988* (New York: I. B. Tauris & Co Ltd., 1996)

O'Ballance, E., *The Gulf War* (London: Brassey's Defence Publishers, 1988)

O'Ballance, E., *The Kurdish Revolt 1961–1970* (London: Faber and Faber Ltd., 1973)

Raspletin, Dr. A. A., 'History PVO' (website in Russian: *historykpvo. narod2.ru*), 2013

Richardson, D., *Techniques and Equipment of Electronic Warfare* (London: Salamander Books Ltd., 1985)

Sadik, Brig Gen A. & Cooper, T., 'Les "Mirage" de Baghdad: les Dassault "Mirage" F1 dans la force aérienne irakienne', *Fana de l'Aviation No. 434/2006*; 'Deuxième partie', *Fana de l'Aviation No. 435/2006*

Sadik, Brig Gen A. & Cooper, T., 'Un Falcon 50 lance-missiles: Avion d'affaires contre navire de guerre', *Fana de l'Aviation No. 470* (2007)

Sadik, Brig Gen A., & Cooper, T. *Iraqi Fighters, 1953–2003: Camouflage & Markings* (Houston: Harpia Publishing, 2008)

Salti, P., *The Royal Jordanian Air Force* (Amman: National Press, 2007)

Schmidt, R., *Global Arms Exports to Iraq, 1960-1990* (Santa Monica: RAND, 1991)

Shakibania, M. & Bibak, S., *Tomcat Fights* (TV documentary, Iran, 2012)

Sorby, K. Jr., *Saddam Husayn's Route to the Top in Iraq, 1976–1980* (Bratislava: Institute of Oriental Studies, Slovak Academy of Sciences, 2018)

Stafrace, C., *Arab Air Forces* (Carrolton: Squadron/Signal Publications Inc., 1994)

Svet, O., *Recipient Behaviour in Security Cooperation Relationships: The Use of Military Assistance in the Expansion of the Iraqi Armed Forces, 1968–1990* (London: King's College, 2020)

Sweetman, B., *Soviet Military Aircraft* (London: The Hamlyn Publishing Group Ltd., 1981)

Tripp, C., *A History of Iraq* (Cambridge: Cambridge University Press, 2000)

Woods, K. M., Murray, W., Holaday, T., with Elkhamri, M., *Saddam's War: An Iraqi Military Perspective of the Iran-Iraq War* (Washington D.C.: National Defense University, 2009)

Woods, K. M., Murray, W., Nathan, E. A., Sabara, L., Venegas, A. M., *Saddam's Generals: Perspectives of the Iran-Iraq War* (Alexandria: Institute for Defense Analyses, 2010)

Various volumes of *Armed Forces Magazine* (Egyptian Ministry of Defence, 1950s and 1960s); *El-Djeich* (Algerian Ministry of Defence, 2007-2015); various magazines and journals published by the Iraqi Air Force (Iraqi Ministry of Defence, 1970-1995); *Kanatlar* magazine (Turkey), June 2003.

NOTES

CHAPTER 1

1. I. al-Marashi, & S. Salama, *Iraq's Armed Forces: An Analytical History* (Abingdon: Routledge Middle Eastern Military Studies, 2008) pp. 111–113, 115
2. I. al-Marashi, & S. Salama, *Iraq's Armed Forces: An Analytical History* (Abingdon: Routledge Middle Eastern Military Studies, 2008) pp. 113–114
3. I. al-Marashi, & S. Salama, *Iraq's Armed Forces: An Analytical History* (Abingdon: Routledge Middle Eastern Military Studies, 2008) pp. 116–118. Ultimately, Hardan at-Tikriti was assassinated in Kuwait, in March 1971.
4. Sadik, interview, 03/2006
5. Sadik, interview, 03/2006; Dr. David Nicolle, interview, 04/2002; Gindy, interview, 02/1999; 'Interview with Major General Alwan al-Abossi' (in Arabic), *al-Gardenia*, 4 April 2014. According to Abossi in the same interview, Ibrahim was a graduate of the Pilot Course No. 19 (or 19th Class). This would mean that he graduated from the Air Force College (future Air Force Academy as of the time), in 1957 – which meant that as of 1970 he was a seasoned veteran. However, in the same article Abossi mentions himself as having graduated with the '15th Class, in 1963'. According to that version, Farhad Ibrahim was a novice pilot who graduated with the 19th Class in 1967. This version sounds more plausible, because a large number of IrAF pilots underwent conversion and tactical training for Su-7BMKs in Czechoslovakia between 1968–1971. Furthermore, Abossi, who flew Su-7BMKs at the time, specified that as of 1971, No. 5 Squadron was commanded by Major Salem Sultan Abdullah al-Basu, and that the High Command in Baghdad officially explained Ibrahim's defection as due to the pilot being 'affected by emotional relations to a girl'. Notably, around a year later, Soviet Air Force pilot Yuriy Kilikauskas defected to Iran – but this time flying an Aero L-29 Delfin trainer (see film 342.USAF.50791 NND 61981, created circa 1973 by the 7602nd Air Intelligence Group, Fort Belvoir, VA., available on YouTube.com). Interestingly, some veteran US intelligence officers recalled Kilikauskas flying a much more powerful Sukhoi Su-9.

CHAPTER 2

1. Sadik, interviews, 03/2005, 03/2006, 03/2007 & 10/2007, and information provided by Martin Smisek (based on documentation from the Archive of the Czech Republic, including VÚA-VHA, MNO, 1971, karton 143, sl. 30-3/17, Dodávkové příkazy ročník 1971).
2. Sadik, interview, 03/2005 & 03/2007; Tobchi, interview, 11/2018. Notably, the Tu-22s sold to Iraq were all second-hand aircraft; former Tu-22R reconnaissance bombers of the Soviet air force, refurbished and re-purposed to act as bombers armed with massive conventional bombs such as the FAB-1500, FAB-3000, FAB-5000, and – the biggest design of this kind in the USSR – the 9,000kg (19,841lbs) FAB-9000. According to Schmidt (Global Arms Exports to Iraq, 1960–1990, p. 55), the first Iraqi orders for 15 Mi-6s and 12 Mi-8s were placed in 1972 and 1971, respectively. The second order for 90 Mi-8s followed in 1975.
3. According to research by Martin Smisek, author of the sister mini-series *Czechoslovak Arms Exports*, Iraq acquired a total of more than 6 L-39Cs and nearly 80 L-39ZOs, the last of which were delivered in 1985.
4. Naji, interview, 02/2007; Sadik, interview, 03/2006 & 03/2007
5. Sadik, interview, 03/2006 & 03/2007; Zayyad, interview, 10/2018. The Joint Eastern Command was headquartered in Suweida, in Syria. As of 1968–1970, it was commanded by Major General Azzawi of the IrAF, with Colonel Nimaa Abdullah Dulaymi as Chief of Staff. In addition to the Iraqis, it included officers of the Royal Jordanian Air Force and the Syrian Arab Air Force but was only in irregular contact with the Egyptian Air Force.
6. Sadik, interview, 03/2005. For details on the operations of Tu-16s of the IrAF against Israel in June 1967 Arab-Israeli War, see Volume 1.
7. Naii, interview, 02/2007 & Sadik, interview, 03/2006 & 03/2007
8. Naji, interview, 02/2007 & Okasha, interview, 07/2011
9. Naji, interview, 02/2007. Notably, most of Iraqi sources quote the rank of Wallid Abdul Latif as-Samarrai as 'Major'. Actually, this was the rank assigned to him posthumously.
10. Ali Tobchi, interview, 11/2013
11. Naimi, intervew, 02/2007
12. Abossi, interview, 07/2014 & Sadik, interviews, 03/2005 & 03/2007
13. Ali Tobchi, interview, 08/2013
14. Sadik, interview, 03/2005 & *IraqiAirForceMemorial.com*
15. Sadik, interview, 03/2005 & *IraqiAirForceMemorial.com*
16. Sadik, interviews, 03/2005 & 03/2007; Abossi, interview, 07/2014. For comparison, one of Jinkir's colleagues, Osama al-Mahdawi, recalled him as a falling in the line of duty while serving on Su-7BMKs with No. 1 Squadron, but flying a MiG-17PF (colloquially known as the 'MiG-17bis' in the IrAF). Based on experience with other of their collections, and the documentation in Sadik's possession, the authors are convinced that Sadik and Abossi are correct in this regard.
17. Sadik, interviews, 03/2005 & 03/2006; al-Awsi, interview, 02/2006; Bakir, interview, 02/2007; Ali Tobchi, interview, 02/2015. It is possible that the Skyhawk in question was one of the TA-4 two-seaters deployed by the IDF/AF from around this time to report the activities of Syrian SAM sites.
18. Naji, interview, 02/2007; Okasha, interview, 07/2011; additional details provided by IDF-veteran and historian, Efim Sandler.
19. Sadik, interviews, 03/2007 & 10/2007
20. Unless stated otherwise, based on Sadik's interviews, 03/2005 & 03/2006.
21. Notably, by this time the first 29 Mil Mi-8 helicopters were already in Iraq, but their crews were still undergoing conversion training. In the rush to support the war effort, conversion training was interrupted, and all available pilots and navigators were assigned to fly Mi-4s with No. 2 Squadron.
22. Sadik, interview, 03/2007 & Abossi, interview, 07/2014
23. Sadik, interview, 10/2007. Amongst others, their experiences from the October 1973 War with Israel influenced the Iraqi decision not to purchase any of the then 'most modern' 2K12 Kub/Kvadrat (ASCC/NATO-codename 'SA-6 Gainful') SAM systems for the IrAF. Instead, only enough of these were acquired to equip four air defence brigades of the Army: the 155th, 162nd, 175th and 185th Missile Brigades, assigned to the 3rd, 6th, 10th, and 12th Armoured Divisions, respectively.
24. Sadik, interviews, 03/2005 & 03/2006; Abossi, interview, 07/2014

CHAPTER 3

1. Sadik, interviews, 03/2005, 03/2006, 03/2007; Trip, pp. 210-211
2. Sadik, interviews, 03/2005, 03/2006, 03/2007; Flintham, p. 180
3. Notably, No. 1 Squadron completed its conversion to Su-20s in a great hurry and went into combat against Barzani's insurgents in April 1974. A month later, its pilots began receiving training in

nocturnal operations. During one related sortie, the jet piloted by 1st Lieutenant Hassan Obeid Kazem – a veteran of the October 1973 Arab-Israeli War, when he flew Su-7BMKs with No. 5 Squadron – crashed near Baba Karkar, killing its pilot.

4 *Engine-Related Problems with Su-20s* (see Bibliography for details); Sadik, interviews, 03/2005, 03/2006, 03/2007 & Iraqi Air Force Memorial website. Al-Obeidi's body was found only following the Algiers Accords of 1975. He was buried in Tal Afar.

5 Sadik, interview, 03/2006; Major General Fawzi al-Barzanji, 'Exploits of the valiant Iraqi Army, 47 Years ago' (in Arabic), Facebook.com, 28 October 2021; *IraqiAirForceMemorial.com*. Wounded in leg when his aircraft was hit, Shallal – graduate of the 19th Class of the Air Force Academy – ejected safely and was captured by Barzani's Kurds. He refused to cooperate under interrogation and was returned to Baghdad. He continued serving as instructor at the Air Force Academy until retirement.

6 Sadik, interviews, 03/2005, 03/2006, 03/2007 & DIA, *Electronic Warfare Forces Study – Iraq*, (henceforth *EWFS-Iraq*)

7 Sadik, interview, 10/2007

8 C. Tripp, *A History of Iraq* (Cambridge: Cambridge University Press, 2000), p. 213

9 Summary based on all available accounts.

CHAPTER 4

1 Marashi *et al., op. cit,* pp. 122–125

2 Unless stated otherwise, based on Sadik, interviews, 03/2006, 03/2007, 10/2007 & Ljiljana Klepo (former HR Manager of Vranica, Sarajevo), interview, 02/1996. According to Klepo, most of the bases described in this sub-chapter were designed by a team of engineers hired by the company IBT, from Belgrade, especially for the purpose of cooperation with Baghdad. The engineers in question were not only involved in constructing 'Objekt Klek' – the underground element of Bihac AB – but had received special permission (from Belgrade, and from Baghdad) to tour all the future construction sites in Iraq.

3 For details on the facility in question, see Dimitrijevic, *Tito's Underground Air Base*

4 Unless stated otherwise, based on Sadik, interviews, 03/2006 & 03/2007

5 According to Schmidt (*Global Arms Exports to Iraq, 1960–1990*, p. 57), Iraq acquired no less than 840 'AA-2 Atoll' missiles in 1975 alone.

6 Relatively little is known about earlier career of Major Muhammad Mudher al-Farhan, except that he flew Su-7s and was stationed at al-Qut at earlier times. By 1980, he would serve as the CO of No. 29 Squadron. Ultimately, Farhan reached the rank of a Major General and was assigned the CO Intelligence Department.

7 While no firm confirmation is available in this regard, the authors are assuming that the Soviets actually delivered 38 refurbished MiG-23Bs to Iraq: these were aircraft from the first series manufactured for the Soviet Air Force in the early 1970s in order to satisfy demands of the GenStab – which was meanwhile desperate to introduce to service a type that experienced repeated delays in research and development.

8 Unless stated otherwise, based on Sadik, interviews, 03/2005, 03/2006, 03/2007

9 Raspletin, *History PVO website* (historykpvo.narod2.ru), 2013 & Sadik, interviews, 03/2006 & 03/2007

10 Unless stated otherwise, based on Sadik, *Iraqi Air Force Electronic Warfare during the Iran-Iraq War, 1980–1988*

11 Hoyt, (page number not declared) & CIA, *Iraq's Intelligence Services: Regime Strategic Intent – Annex B*, 23 April 2007, CIA, FOIA Electronic Reading Room (henceforth 'CIA/FOIA/ERR').

12 Sadik, 03/2006 & Hoyt. Because of his military expertise, Amer Rasheed was often referred to as 'Missile Man'.

CHAPTER 5

1 Recollections of veteran IrAF officers and foreign officers that came into contact with him – such as the pilots of the Royal Jordanian Air Force during the June 1967 Arab-Israeli War – about Sha'ban, could not be more contradictory. Ironically, Brigadier General Ahmad Sadik Rushdie al-Astrabadi, a Shi'a, could be described as something like 'Sha'ban's greatest fan': he praised him as very aggressive, offensive and a high-technology-oriented commander, and an excellent administrator and organiser. Sadik declared Saddam Hussein's decision to appoint Sha'ban in 1976, and then to reactivate and return him to service for a second tour of duty as Chief of Staff in 1984, as 'the crucial decision' of the Iran-Iraq War. On the contrary, Major General Abossi, a Sunni, described Sha'ban as 'pretty weak' as a pilot, 'not effective' as Chief of Staff IrAF, and somebody who was 'not admired in the air force as much as his predecessors' nor successors.

2 For details on Riyadh and the flow of developments that resulted in the idea for the IADS in Egypt, see Cooper *et al., MiGs in the Middle East, Volume 2*.

3 Sadik, interview, 10/2007

4 Sadik, interviews, 03/2005, 03/2006, 10/2007; Fukuyama, p. 52; Schmidt, *Global Arms Exports to Iraq, 1960–1990*, pp. 57–58 & Adeed I. Dawisha, 'Iraq: The West's Opportunity', *Foreign Policy*, No. 41, pp. 134–155. Notably, Sadik stressed that the IrAF continued asking for British conditions for a sale of F-4Ks until at least 1983–84.

5 Irrespective of what French company was delivering what part of the aircraft, Thales acted as the official French representative. Thus, juridically, Avions Marcel Dassault never sold any Mirages to Baghdad, and the SNECMA company never delivered any Atar engines for them: it was Thales that did so. For details, see Cooper *et al., Iraqi Mirages*.

6 Sadik, interviews, 03/2005, 03/2006 & Schmidt, *Global Arms Exports to Iraq, 1960–1990*, pp. 57–58. For a detailed description of Project Baz-AR, see Cooper *et al., Iraqi Mirages*.

7 Retired DIA analyst, interviews provided on condition of anonymity, 02/2001 & 10/2002

8 Unless stated otherwise, based on Sadik, interview, 10/2007

9 Unknown to the Iraqis, since 1970, Crypto AG – a company which specialised in communications and information security that eventually sold its equipment to more than 120 countries around the world (and well into the twenty-first century) – was secretly owned by the CIA and the Bundesnachrichtendienst (BND) of West Germany. It not only ran smear campaigns against all the possible rivals on the market, or plied government officials with bribes, but also trained all the users of the equipment it was manufacturing. Obviously, this enabled the CIA, the National Security Agency of the USA (NSA) and the BND to easily break the codes and listen to the most sensitive communications of all the countries that used encrypting machinery made by Crypto AG. Primarily thanks to the work on the Kari IADS, Iraq became its second largest customer in the Middle East, right after Iran. The IrAF is known to have bought telephones, faxes, and related encryption systems made by Crypto AG. For details, see Greg Miller, 'The Intelligence Coup of the Century', *The Washington Post*, 11 February 2020.

10 When one considers the idea for the Kari IADS/ATMS, it should be kept in mind that it depended on the French computer technology of the late 1970s, and even older Soviet computer technology (installed in the Bastion systems, and the Azurk-1ME, which apparently served as a back-up to the Kari's main computer, made by Thales). What Thales did was to use the Bastion systems Iraq acquired from the USSR as centrepieces for the command posts of ADS, Sector Operation Centres, and ground control stations of Kari; their computers prepared the 'radar picture' of the situation before it was forwarded to the main computer of the Kari IADS/ATMS in Baghdad, in turn decreasing its burden, shortening the reaction time, and creating redundancy in the case of combat damage or technical failure. To a certain degree, it can be said that Kari included a very early version of a 'multi-processor unit'.

11 Notably, the IrAF had its own ground-based air defence units equipped with a wide range of anti-aircraft artillery pieces, ranging from 12.7mm and 14.5mm heavy machine guns, to 23mm, 37mm and 57mm automatic guns. Indeed, every single air base was protected by an air defence battalion including three or four batteries of 14.5mm ZPU-4 quadruple machine guns, and a battery of towed 23mm ZU-23 guns. Contrary to many Western reports, the IrAF never operated any 85mm anti-aircraft guns, while its old pieces of 100mm calibre were all withdrawn from service and stored by 1978 (Sadik, interview, 10/2007).

12 Unless stated otherwise, based on Sadik, interviews, 03/2006, 03/2007, 10/2007 & Klepo, interview, 02/1996. According to Klepo, the total value of the contract for the original Project 202 was over US$1 billion. While the value of the second contract remains unknown, the third and final air base construction contract between Baghdad and Belgrade – including the one for Project 1100 – was worth another US$4.3 billion. Even if not all the resulting facilities were completed by the time, most were handed over to the IrAF by 1987.

13 Sadik, interview, 10/2007

14 Barbouti, interview to Ali Tobchi, 08/2018

CHAPTER 6

1 Based on Svet, p. 350; FBIS-MEA-77-140, 1977; Fukuyama, pp. 56–57; Marashi *et al.*, p. 127; Timmerman, pp. 74–122; Adeed I. Dawisha, 'Iraq: The West's Opportunity', *Foreign Policy*, No. 41, pp. 134–155; Tripp, pp. 217–221.

2 Ironically, most of the money in question was then spent for the construction of a new, 950 kilometre-long (589 miles) pipeline from Kirkuk to Turkey. However, this statement did make Moscow forget about Barzan's Kurds for a while; the Kremlin interpreted it as a sign of good will.

3 Sadik, interview, 10/2007; Fukuyama, pp. 59–62, 72; Schmidt, *Global Arms Exports to Iraq, 1960–1990*, pp. 53; DIA, *Electronic Warfare Forces Study – Iraq*, 9 August 1990, CIA/FOIA/ERR. Notably, based on information available in contemporary publications, some of Schmidt's data about numbers of aircraft ordered – and 'delivered' – to Iraq were off. Actually, the IrAF began ordering new combat aircraft from the USSR 'by 18', because that was the officially authorised number of aircraft per squadron of tactical fighters (bomber and transport units were authorised different complements). The figure was based on the premise that each unit was to consist of four flights of four aircraft and keep two jets as attrition reserve. Obviously, this fact was entirely unknown in the West.

4 Unless stated otherwise, based on Marashi *et al.*, pp. 127–128; Tripp, pp. 218–219; Marvine Howe, 'New Anti-Communist Actions were reported – and denied – last Week', *The New York Times*, 7 January, 1979 & articles about Iraq in 1978 and 1979 from *Collier's Encyclopedia Yearbook 1979* and *Collier's Encyclopedia Yearbook 1980*.

5 Sadik, interview, 10/2007; Tripp, p. 217

6 K. Sorby, Jr., *Saddam Husayn's Route to the Top in Iraq, 1976–1980* (Bratislava: Institute of Oriental Studies, Slovak Academy of Sciences, 2018) pp. 397–398

7 Fawzy al-Barzanji, 'Some History: What marked the Iran-Iraq War' (in Arabic), *al-Gardenia*, 26 May 2016 (Barzanji served as liaison officer at the Ali Ibn Abu Talib AB of the early 1980s, and interviewed Major General Kamal Abdul Sattar al-Barzanji for this article), and Sadik, interview, 10/2007

8 Abossi, interivew, 07/2014

9 Sorby, p. 401 & Ranj Alaadin, 'Iraq's failed Uprising after the 1979 Iranian Revolution', *Brookings*, 11 March 2019

10 Sadik, 03/2005; Fawzy al-Barzanji, 'Some History: What marked the Iran-Iraq War' (in Arabic), *al-Gardenia*, 26 May 2016; CIA, *Iran-Iraq: Determining Who Started the Iran-Iraq War*, 25 November 1987, CIA/FOIA/ERR. According to Sadik and the article in *al-Gardenia*, by August 1980, the authorities in Baghdad had registered over 240 violations of airspace and 493 ground attacks on Iraq, all by Iranian armed forces, and issued 293 diplomatic protests to Tehran.

11 For a detailed description of the operations in question, see Hooton *et al.*, *Iran-Iraq War, Volume 4* (details in Bibliography).

12 For details on related Iraqi preparations, see Hooton *et al*, *Iran-Iraq War, Volume 1*.

13 Unless stated otherwise, based on Sadik, *Iraqi Air Force Electronic Warfare during the Iran-Iraq War, 1980–1988*

14 Sadik, interview, 10/2007; Chad E Nelson, 'Revolution and War: Saddam's Decision to Invade Iran'; *Middle East Journal*, Vol. 72, No. 2, 2018 & Tehran Domestic Service, 'Tehran reports further on Iraq Air Attack', 7 June 1979, FBIS-MEA, Vol. V, No. 111, p. R1.

15 Sadik, interview, 10/2007

16 K. Sorby, Jr., *Saddam Husayn's Route to the Top in Iraq, 1976–1980* (Bratislava: Institute of Oriental Studies, Slovak Academy of Sciences, 2018), pp. 401–402; Woods *et al.*, *Saddam's War*, p.32 & Sadik, interview, 10/2007. Precise details of what exactly happened on 4 September 1980 on the border between Iran and Iraq remain elusive. The Iraqi media (for example, *Baghdad Observer*, on 7 September 1980) originally reported a 'severe artillery attack' on – literally – two villages. However, subsequently, this affair was styled as 'the first day of the Iranian aggression', and the Iranians were accused of 'bombing and shelling Iraqi cities and towns', including Baghdad and Basra, and 'killing women and children'. Sufficient to say that there is no evidence that anything of this kind ever happened. Nevertheless, most of the Iraqi sources interviewed over time consider this day – 4 September 1980 – as 'the day the Iran-Iraq War began'.

17 *iraqiairforcememorial.com*

18 IrAF, *An Analytical Study on the Causes of Iraqi Aircraft Attrition during the Iran-Iraq War* (see Bibliography for details) & Sadik, interviews, 03/2006, 03/2007. Ironically, even the above-mentioned IrAF study lists Hamadani as shot down on 14 September, although by this time the Iraqi commission were negotiating with the Iranians about the exchange of prisoners and bodies of those killed during eight years of fighting, and admitted he was lost five days earlier. For this reason, Sadik insisted Hamadani was lost on 14 September 1980, too: having known that pilot personally, the Iraqi Brigadier General continued searching for any kind of information

about the fate of the missing pilot until his arrest and imprisonment by Syrian authorities, on 2 January 2008.
19 Sadik, interview 03/2005. According to Sadik, Sadiq sighted just one F-4E and fired two R-13Ms but could not clearly observe if one or two had hit: this was a typical problem caused by the MiG-21's thick windshield, which severely restricted the pilot's view to the front. In turn, it is perfectly possible that Eskandari's F-4E was first hit by one of Sadiq's R-13Ms, and then felled by friendly ZU-23s and machine gun fire. Indeed, such a 'combination' would best explain why he ejected shortly after crossing the border.
20 According to Fawzy al-Barzanji (in 'Some History'), on 17 September 1980 Major Kamal Abdul Sattar al-Barzanji shot down the F-5E piloted by Hossein Ali Reza Lashgari over the town of Baqubah, 'in the outskirts of Baghdad'. Baquabah is about 50km/31 miles northeast of Baghdad, but Lashgari was actually shot down by a ZSU-23-4 Shilka self-propelled anti-aircraft gun of the Iraqi Army in the Fakkeh area, on the border between Iran and Iraq, on 18 September 1980. Tragically for him, he became the first Iranian prisoner of war in Iraq. On Saddam's order, he was held imprisoned until 1998 – for a total of 18 years or 6,410 days – supposedly as 'evidence of the Iranian aggression', and that Iran 'started the war'.

CHAPTER 7

1 O. Svet, *Recipient Behaviour in Security Cooperation Relationships: The Use of Military Assistance in the Expansion of the Iraqi Armed Forces, 1968–1990* (London: King's College, 2020) pp. 148–149
2 Sadik, interview, 10/2007
3 Sadik, interview, 03/2006. The decision to deploy the MiG-23MS of No. 39 Squadron is heavily criticised by IrAF veterans until this very day. The reason for it was an engine-related grounding of the entire Su-20 fleet of No. 1 Squadron in the summer of 1980, described below.
4 Unless stated otherwise, based on Sadik, interview, 10/2007 & Tobchi, interview, April 2017. Figures for deliveries of air-to-air missiles are based on Schmidt, *Global Arms Exports to Iraq, 1960–1990*, p. 53, with minor corrections according to data supplied by Sadik and Tobchi.
5 Sadik, interviews, 03/2005 & 03/2006; Tobchi, interview, 03/2015. Notably, Sadik recalled '84' as the designation of the first MiG-25 unit, causing some confusion between the subsequently established No. 84 Squadron, which operated Messerschmitt-Bölkow-Böhm MBB.105 helicopters, and No. 87 Squadron, which was the first unit officially declared operational on MiG-25s, in 1981.
6 According to Sadik (interview, 03/2006), the Soviets delivered no R-23T infra-red homing air-to-air missiles for Iraqi MiG-23MFs.
7 Sadik, *Iraqi Air Force Electronic Warfare during the Iran-Iraq War*; Tobchi, interview, 04/2017; DIA, *Electronic Warfare Forces Study – Iraq*; CIA Report Z-2016/80, September 1980.
8 Sadik, interview, 03/2005
9 *Engine-Related Problems with Su-20s* (see Bibliography for details). Su-20s of No. 1 Squadron remained grounded until 24 September: after flying a few combat sorties, it was grounded again from 1 until 24 October 1980.
10 Sweetman, *Soviet Military Aircraft*, pp. 63–65 & Sadik, interview, 03/2006
11 Sadik, interview, 03/2006 & Daoud Salman ash-Shwaili, 'The Third Duty' (in Arabic), shrabann.blogspot.com, February 2021
12 Sadik, interview, 10/2007

CHAPTER 8

1 The authors would like to stress that this is how the planning for the Iraqi operation was described to them by several Iraqi sources (foremost Sadik, interviews, 03/2005, 03/2006, 03/2007, 10/2007 & Abossi, 07/2014). Contrary to what has been claimed by dozens of Western commentators and observers since, the same sources have never mentioned any kind of link between the Iraqi planning for the invasion of Iran in September 1980 and the Israeli operations against Arab air forces during the June 1967 War. When directly asked if the latter had any kind of influence, Sadik was astounded. Therefore, and contrary to what is still frequently claimed in Western publications, the IrAF never thought of emulating what the IDF/AF did on 5 June 1967. Indeed, considering how widely spaced and heavily fortified Iranian air bases were, even the Israelis would have had a major problem with trying to emulate their operation from 1967. On the other hand, both Sadik and Abossi stressed, repeatedly, that it was Iran that was 'beating war drums', and solely responsible for the 'pre-emptive' Iraqi reaction. Indeed, Abossi has repeated similar statements in most of his related online publications since; according to him, all accusations of Iraqi aggression are 'fabrications'.
2 Sadik, interview, 10/2007 & Abossi, 09/2010
3 Sadik, interview, 10/2007
4 Sadik, interview, 10/2007
5 Sadik, interviews 03/2005 & 10/2007; IrAF, *Analytical Study*, p.3; Woods, *Saddam's Generals*, pp. 207–210. Not only Sadik, but also several generals interviewed by Woods et al. have stressed that the IrAF received the plans for the attack only on 20–22 September 1980. Some went as far as to describe the plans they had seen as those for a 'training exercise'.
6 Sadik, interview, 03/2007
7 Sadik, interviews, 03/2005, 03/2006, 03/2007 and 10/2007 & Fawzy al-Barzanji, 'Some History: What marked the Iran-Iraq War' (in Arabic), *al-Gardenia*, 26 May 2016.
8 Conclusions based on all above-mentioned interviews, and interviews with more than 50 Egyptian, Syrian, and other Arab veterans over the last 30 years.
9 Sadik, interview, 10/2007 & Abossi, in his write-up *In Memory of the Comprehensive Response*. Sadik never mentioned the total number of sorties flown by the IrAF on 22 September 1980, while Abossi claimed 192. On the basis of currently available information, the conclusion is that the latter figure is perfectly realistic – even if it might include combat air patrols flown by MiG-21s inside Iraqi airspace.
10 Sadik, interview, 10/2007 & Namaki et al., *History of Air Battles, Vol.2*, pp. 270–272
11 Sadik, interview, 10/2007; Namaki et al., *History of Air Battles, Vol.2*, pp. 270–272; Sa'aydoon, interview, 10/2021
12 Sadik, interview, 10/2007 & Namaki et al., *History of Air Battles, Vol.2*, pp. 270–276
13 Fawzy al-Barzanji, 'Some History: What marked the Iran-Iraq War' (in Arabic), *al-Gardenia*, 26 May 2016, claimed that all five MiG-23BNs of Sa'aydon's formation run out of fuel and crashed. Both Abossi (*In Memory of the Comprehensive Response*) and Sa'aydoon (son of Major Hamad Rashid as-Sa'aydoon), denied this. Indeed, as Sa'aydoon reported that his father led the next strike of his squadron during the third wave: this would have been extremely unlikely if the CO of No. 29 Squadron had to eject from an aircraft far away from home-base during the first wave.
14 Sadik, interview, 10/2007 & Namaki et al., *History of Air Battles, Vol.2*, pp. 277–282

15 Sadik (interview, 10/2007), stressed that only two waves were planned – and that both were flown. On the contrary, Abossi (*In Memory of the Comprehensive Response*), stated that the High Command IrAF discontinued the second wave and 'replaced' it with 'individual air strikes based on objectives and achievements of the first wave'.

16 Sadik, interview, 10/2007; Sa'aydoon, interview, 10/2021; Namaki et al., *History of Air Battles, Vol.2*, pp. 292 & iraqiairforcememorial.com. Ironically, all Iraqi sources available nowadays report that both the Sukhoi and the MiG formations underway to Orumiyeh failed to reach their targets, while the Iranians stress that the runway of Rezaie was heavily hit on 22 September 1980.

17 Sadik, interview, 10/2007 & Dr. Alwan al-Abossi, 'We embrace the Sky, Part 2, C5' (in Arabic), *Al-Gardenia*, 21 February 2017. Sadik recalled 1st Lieutenant Ra'ad Hameed as killed in action. According to Abossi, the young pilot survived 10 years in Iranian captivity: he was exchanged in 1990 and was promoted to the rank of a Colonel on return to Iraq.

18 Boeing 707-3J9C was, essentially, an export variant of the Boeing KC-135A tanker.

19 A. R. Namaki, Brigadier General, Khalili, Brigadier General H., & Djavaheri, Colonel. A. R., *History of Air Battles, Volume 2: Invasion and First Response* (in Farsi), (Tehran: Centre of Strategic Studies, IRIAF, 2016) pp. 275–284

20 A. R. Namaki, Brigadier General, Khalili, Brigadier General H., & Djavaheri, Colonel. A. R., *History of Air Battles, Volume 2: Invasion and First Response* (in Farsi), (Tehran: Centre of Strategic Studies, IRIAF, 2016) pp. 319–342. The IRIAF F-4E was repaired after the war. Similarly, the Boeing 707-3J9C hit by bombs dropped by Tu-22s was initially written off. Years later, the IRIAF assembled a large team of technicians that completely rebuilt its right wing and returned it to service. The civilian B707 was converted into a restaurant – which is still open!

21 A. R. Namaki, Brigadier General, Khalili, Brigadier General H., & Djavaheri, Colonel. A. R., *History of Air Battles, Volume 2: Invasion and First Response* (in Farsi), (Tehran: Centre of Strategic Studies, IRIAF, 2016), p. 340

22 Sadik, interview, 10/2007

23 A. R. Namaki, Brigadier General, Khalili, Brigadier General H., & Djavaheri, Colonel. A. R., *History of Air Battles, Volume 2: Invasion and First Response* (in Farsi), (Tehran: Centre of Strategic Studies, IRIAF, 2016) pp. 265–269

24 Sadik, interview, 10/2007

25 Tobchi, interview, 08/2019. Graduate of the 12th Class Air Force Academy, Sabbah served with No. 6 Squadron before being reassigned to the Air Force Academy, in the mid-1970s. He was reassigned to the FLS during preparations for the Iraqi invasion. Latest claims according to which Wallid and Sabbah 'destroyed four F-4 Phantoms on the ground' in Eslamabad-e-Gharb, making circles in the Iraqi social media, are within the realms of fantasy; the IRIAF was not only not in condition to start dispersing its aircraft before the Iraqi invasion, but would never deploy any to such an exposed site as Elsamabad-e-Gharb. Finally, the Iranians are known to have lost only two F-4Es on 22 September 1980: one knocked out by Tu-22s at TFB.1, and one shot down over Abu Ubayda Ibn al-Jarrah AB, as described below.

26 A. R. Namaki, Brigadier General, Khalili, Brigadier General H., & Djavaheri, Colonel. A. R., *History of Air Battles, Volume 2: Invasion and First Response* (in Farsi), (Tehran: Centre of Strategic Studies, IRIAF, 2016) p. 296. The downed Number 2 from this air strike, the F-4E piloted by Salehi and Heydari, wore the IRIAF serial 3-5685.

27 A. R. Namaki, Brigadier General, Khalili, Brigadier General H., & Djavaheri, Colonel. A. R., *History of Air Battles, Volume 2: Invasion and First Response* (in Farsi), (Tehran: Centre of Strategic Studies, IRIAF, 2016) p. 297; Sadik, interview, 10/2007; retired DIA analyst, interview, 10/2002.

28 Huge orders for new aircraft and helicopters issued between 1970 and 1980 caused quite a disorder in the usual IrAF serialling system, which – at least nominally – still depended on the order in which the aircraft actually reached Iraq. While the Quartermaster General (Deputy Logistics) attempted to issue serial numbers in batches, several cases are known where two aircraft of different types, or an aircraft and a helicopter delivered around the same time, received the same serial number. This was mainly the case where serials were applied on aircraft or helicopters already in Czechoslovakia, France, or the USSR, before their actual shipping to Iraq. Furthermore, the mass of helicopters listed in this table were reassigned to the IrAAC in the summer of 1980.

ABOUT THE AUTHORS

TOM COOPER
Tom Cooper is an Austrian aerial warfare analyst and historian. Following a career in the worldwide transportation business – during which he established a network of contacts in the Middle East and Africa – he moved into narrow-focus analysis and writing on small, little-known air forces and conflicts, about which he has collected extensive archives. This has resulted in specialisation in such Middle Eastern air forces as of those of Egypt, Iran, Iraq and Syria, and various African and Asian air forces. In addition to authoring and co-authoring more than 50 books – including an in-depth analysis of major Arab air forces during the wars with Israel in 1955–1973 – and over 1,000 articles, Cooper is working as editor of Helion's five @War book-series.

MILOS SIPOS
Milos Sipos is a Slovakian military historian. While pursuing a career in law he has collected extensive documentation on interconnected political, industrial, human resources and military-related affair in Iran, Iraq and Syria. His core interest is a systematic approach to studies of their deep impacts upon combat efficiency and the general performance of local militaries. After more than 10 years of related work on the ACIG.info forum, he specialised in research about the Iraqi Air Force and the Syrian Air Force, and co-authored the books *Iraqi Mirages* and Volume 1 of *Wings of Iraq* published in Helion's @War series.